MAR 0 6 2018

A Magical World

A Magical World

Superstition and Science from the Renaissance to the Enlightenment

DEREK K. WILSON

PEGASUS BOOKS
NEW YORK LONDON

A MAGICAL WORLD

Pegasus Books, Ltd.
148 West 37th Street, 13th Floor
New York, NY 10018

Copyright © 2018 by Derek Wilson

First Pegasus Books hardcover edition February 2018

All rights reserved. No part of this book may be reproduced in whole or in part without written permission from the publisher, except by reviewers who may quote brief excerpts in connection with a review in a newspaper, magazine, or electronic publication; nor may any part of this book be reproduced, stored in a retrieval system, or transmitted in any form or by any means electronic, mechanical, photocopying, recording, or other, without written permission from the publisher.

ISBN: 978-1-68177-645-3

10 9 8 7 6 5 4 3 2 1

Printed in the United States of America
Distributed by W. W. Norton & Company, Inc.

Contents

INTRODUCTION

‘42’

Douglas Adams, in his series of spoof sci-fi novels *The Hitchhiker's Guide to the Galaxy* (1979–1992), has a supercomputer provide the answer to the question of ‘life, the universe and everything’: ‘42’. When the galactic voyagers fail to understand this answer, they are told that the reason is that they do not understand the question. The author was inundated with queries, calculations and observations from readers intrigued to unravel the cosmic significance of this magic number. Of course, as he pointed out over and over again, there was no significance. 42 was a figure plucked from the air. It was a *joke*.

But a joke is not funny if it does not relate to recognisable human experience. The fact is that from some point in the early history of *Homo sapiens* (‘reasoning man’) we have been asking ourselves, ‘What's it all about?’; ‘What on Earth (literally) are we doing stuck on this speck of rock spinning in space?’

Broadly speaking, we have always responded in two ways to the mystery of being: we have explored nature and supernature. The first method of inquiry we call ‘science’, an intellectual discipline that restricts itself to the study of the material universe. The second is much more difficult to define. It embraces religion, philosophy and the arts and is based on the assumption that man is not confined by materiality; that he is possessed of a soul and is related to another sphere of existence which is above and beyond what we experience with our five senses. Both of these are responses to the problems of ‘life, the universe and everything’ and they are *not* mutually exclusive. If they were it would be impossible for anyone to be a dedicated scientist and, at the same time, sincerely religious.

We can, of course, choose to regard them as incompatible. The Christian fundamentalist may insist that the theory of evolution contradicts biblical revelation about the Creator and is, therefore, wrong. The atheist fundamentalist may assert that nothing whatsoever lies beyond the material universe. Both of these are faith statements, quite incapable of proof. If we set them aside, we discover between the two extremist positions a vivid and multi-coloured intellectual tapestry which forms an important part in the history of our civilisation. In the following pages we will explore one vital 'panel' of that tapestry – the world of changing ideas that existed between the years 1450 and 1750; between the Renaissance and the Enlightenment.

So far, I have said nothing about 'superstition'. There are two reasons for this: the word is difficult (perhaps impossible) to define and it is emotionally loaded. No dictionary that I have consulted offers a wholly satisfactory definition of the word. The *Oxford English Dictionary*'s headline explanation is: 'Unreasoning awe or fear of something unknown, mysterious or imaginary, esp. in connection with religion'. Merriam-Webster gives us: 'A belief or way of believing based on fear of the unknown, any faith in magic or luck'. Google offers: 'Excessively credulous belief in or reverence for the supernatural'.

I could go on but the result would only be more confusion. Who decides what degree of belief is 'excessive'? Why is it 'unreasoning' to be in awe of the immensity and complexity of a universe we are scarcely beginning to understand? And is it not questionable to lump together 'magic', 'luck', 'mystery' and 'religion'? Irrational phobia of the number thirteen is scarcely in the same league as acceptance of the theory of final causes – the idea that everything that exists does so for a specific purpose – debated by philosophers, theologians and scientists from Aristotle to Dawkins. Throughout history all manner of beliefs have been dismissed as 'superstitious'. Plato applied the term to fear of the Greek gods. St Paul found the very multiplicity of deities worshipped in Athens superstitious. Luther rejected as superstition certain practices officially sanctioned by the Pope. David Hume believed that superstition was the outworking of 'certain unaccountable terrors and apprehensions' inherent in the human mind. It seems that 'superstition' is simply a pejorative term used by anyone to describe beliefs or practices to which he/she does not happen to subscribe. Since we need reasonable precision as the basis for our study, I am going to stick my neck out and offer my own definition of 'superstition'. It is something, I suggest, that we all experience to a greater or lesser degree, our psychological response to the unknown, a response that appropriately should comprise intelligence, humility and respect.

The historian's task is not to promote belief or disbelief, or to define 'truth'. What I shall be attempting in the following pages is to record the activities and opinions of some of the great thinkers who contributed to the debate about 'life, the universe and everything'. It would, however, be unfair if I did not come clean about two of my own preconceptions. The first is a rejection of the concept of 'progress'. To assert that we are on an upward journey from ignorance to enlightenment, savagery to civilisation, darkness to light is simply to fly in the face of what our own common sense reveals about contemporary humanity. When 80 per cent of the world's population live below the poverty line we can scarcely claim to have 'peaked' in terms of progress. To conquer killer diseases or place probes on distant planets are phenomenal achievements but they do not make us 'better' people. It is not the discovery of new knowledge that provides a basis for moral judgements about society. What matters is the use we make of that knowledge. Every age has to be judged by its own criteria. That is why I shall not simply be treading the well-worn path 'from Petrarch to Hume', which represents the great thinkers of the period as intellectual milestones on a pilgrimage of inevitable human progress 'from Dark Ages to Enlightenment'.

Human history is not so obligingly simple. This book is called *Superstition AND Science*, not *Superstition TO Science*, because in all ages the two are interwoven.

My second preconception is that we must see the great intellects – whose discoveries and inventions we rightly celebrate – not as cloistered solitaries, immune to what was happening beyond the protective walls of their own genius. They did not exist in a parallel universe. They were in and of the world. They were affected by the same tumults, tragedies and triumphs that shaped the lives of their contemporaries. Their discoveries and inventions were, in part, responses to the issues of the day. If we do not see John Dee, Galileo, Descartes, Newton and the rest as men of their times we do not see them at all. And we must see them warts and all; their failures as well as their successes – the independence of thought that led to enlightenment and the blinkered vision that restricted their accomplishments – in effect, the entanglement of superstition and science in all their reasoning. The pioneer thinkers in the field of what Francis Bacon would call 'natural philosophy' are those who, along with artists, theologians, musicians and all liberated thinkers, have helped us to explore the complexities of our shared humanity and the cosmos to which each of us pays a fleeting visit. I hope that their stories will help us to arrive at more satisfying answers to the big questions than '42'.

CHAPTER ONE

New wine and old wineskins

Medieval Europe=Western Christendom. The culture of what we rather foolishly call the European 'continent' was Christian. What defined the western chunk of the Euro-Asian landmass was not its physical geography but its adherence to a despised minority religion founded in Palestine which became the official creed of the Roman Empire in the early fourth century A.D. For the next millennium Christendom had two power centres: Rome and Constantinople. Then, in 1453, the eastern capital fell to the invading Muslim army. Thereafter the history of 'East' and 'West', roughly divided by the valleys of the Danube and the Vistula, diverged radically, one region being contested by the Ottoman and Muscovite empires and the other blighted by conflict between powerful political regimes ruled from Madrid, Vienna and Paris. The only 'misfit' in this geographical/political/cultural map was the ancient Muslim enclave of Granada in southern Spain and, by 1500, that had been overrun by the Christian *Reconquista*.

Culturally, all roads in the West led to Rome. It was not only the religious life of Europe that was unified through networks of dioceses, parishes and monasteries; science was also the preserve of the Church. Yes, *science*. Despite historian Edward Gibbon's assertion that the Church supplanted 'in an unnecessarily destructive way the great culture that preceded it', pursuit of knowledge in all its forms continued in abbeys, royal courts and, later, universities throughout what became known as the Dark Ages and the Middle Ages. It was the fourteenth-century poet Petrarch who coined the term 'Dark Ages' as a means of dismissing the debased Latin of the preceding centuries, but it was given wider meaning by post-Enlightenment scholars to identify what they considered to be a sunless chasm of ignorance and barbarity lying between the fertile intellectual uplands of the late Classical and the Renaissance eras. The truth is much more complex. Stagnation there certainly was. This was, in part, due to the frequent dislocation of civil society. But there was also intellectual vitality, interest in the wisdom of the past and discovery of fresh wisdom. Though now rejected by scholars, the term 'Dark Ages' still occasionally raises its head, so it might be as well to get it out of the way, once and for all.

The Roman Empire died like an exhausted predatory beast whose resources were, finally, inadequate for the task of overpowering all the rivals by which it was surrounded. Those rivals were also engaged in their own territorial conflicts, which continued after the last legions had withdrawn. They lacked the technical know-how and also the will to maintain the relics of their conquered foe – the roads, aqueducts, villas and temples, as well as the political, religious and social institutions of which those structures were the sloughed-off skin.

But by no means did that signify that the northern tribes were 'uncultured'. They have left plenty of evidence of creativity and ingenuity. Numerous 'treasures' retrieved by archaeologists at widespread sites indicate high levels of artistic achievement. Nor was it only in matters of personal adornment for the wealthy that Dark Age technology excelled. Among the inventions of the post-Roman world were the wheeled plough and the water-driven mill. (By 1086, when William the Conqueror commissioned the Domesday survey, there were 5,624 such mills operating throughout England). It has sometimes been asserted that the decay of Roman roads proves the backwardness and insularity of European communities, but if such 'highways' were abandoned it was for good, practical reasons. Imperial roads were built for marching men and existed to facilitate conquest and political control. They were not made, nor were they suitable, for civilian trade and transport. They were narrow, stone-topped and not well-suited for the wagons that played an increasingly important part in Dark Age economies.

It is significant that one of the prominent figures in northern mythologies was Völund or Wayland, the smith. This semi-divine hero was the subject of numerous Nordic and Germanic legends. It was believed that he could fly and manifest other magical properties. Swords forged by him were invincible. His helmets and chainmail had especially protective properties. In England it was said that if a horseman left his mount with some money outside Wayland's Smithy, a long barrow in Oxfordshire, he would return to find the animal expertly shod. Such tales are an early example of supernatural gifts being associated with technical skill. Ironworkers were so important to their communities, their craft of fashioning weapons and tools from crude metal so mysterious and the noise and fire of their workshops so impressive, that it is not surprising that they should have been thought to possess superhuman attributes. Blacksmiths not only provided conventional weapons, armour and farm implements; they invented new ways of doing things. For example, the 'simple' development of the iron horseshoe brought about enormous changes in everyday life. Once a suitable harness had been invented, horses replaced oxen as draft animals. A horse could plough a field twice as quickly as an ox. New transport possibilities led, in turn, to innovations in the construction of wagons. Wheel brakes and pivoting front axles were just two of the improvements made by the application of the smith's 'science'.

We might explore other changes to everyday life which demonstrated the ingenuity and ever-developing techniques of Dark Age craftsmen: three-field agriculture employing crop rotation, fish-farming and the building of ships suitable for northern waters. When the inheritors of Roman buildings demolished them it was not because they were too ignorant or too resentful to make use of them. It was because they could make *better* use of the stone – for bridges, walls, churches and manor houses for the leaders of local society.

Even warfare benefitted from the development of new technology. In 732 Charles Martel, King of the Franks, led the first recorded charge of mounted knights, a formidable fighting force made possible by the development of stirrups and high-cantle saddles. During the post-Roman centuries European society did not descend into a slough of barbarism; it *evolved*, developing its own cultural dynamic.

When we turn from technology to the realm of aesthetics – to matters of the soul, rather than the body – to art, music, intellectual enquiry, and the pursuit of *scientia* – we find ourselves in that remarkable institution, the Christian Church. In the fading empire the men who inherited the leadership of society – emperors, praetors, quaestors, tribunes and aediles and other officials – were Christians. Their beliefs and ethics shaped the society of the later Roman world, gradually absorbing old religions and philosophies. The official religion of the post-Constantine empire was robust. It was clear about its basic tenets but also able to adjust to changing circumstances in a turbulent world. Late Roman Christianity was, within well-defined boundaries, adaptable while the *raisons d'être* of the empire itself were not, and that, in large measure, explains why the faith of the Nazarene Carpenter prevailed. Victorian poet Algernon Charles Swinburne accurately observed it – and deplored it: 'Thou hast conquered, O pale Galilean, the world has grown grey from thy breath'. Yet even his nostalgic paganism acknowledged the revolutionary new ethic of Christianity:

> O Gods dethroned and deceased, cast forth, wiped out in a day!
> From your wrath is the world released, redeemed from your chains, men say.

New Gods are crowned in the city: their flowers have broken your rods;
They are merciful, clothed with pity, the young compassionate Gods.

Algernon Swinburne, 'Hymn to Proserpine'

But that is only part and scarcely half of the picture. The importance of the Christian Church lay as much in what it preserved as in what it destroyed. As the inadequacies of the old order became progressively more apparent, the new faith was able to 'deal with the sterility or incoherence produced by its predecessor, account for the previous difficulty in doing so, and carry out these tasks "in a way which exhibits some fundamental continuity of the new . . . structures with the shared beliefs in terms of which the traditions of inquiry had been defined up to that point"'.*

The administrators, churchmen and scholars of Europe in transition were reared in the literature of the classical world and maintained a respect and admiration for the philosophers, poets and historians of Greece and Rome. It was in the monasteries of the Dark Ages that scholars collected and copied old manuscripts for their libraries. Their collections were not comprehensive but they did preserve several important philosophical works. In doing so they pondered the relationship of the classical greats to the Bible and to their own theologians. This involved them in juggling three intellectual balls. They had revelation in the shape of their sacred scriptures. From the ancient masters they had philosophy based on speculation. And those same masters offered science supported by empirical observation.

It was Plato (*c.*427–*c.*347 BC) and his pupil, Aristotle (384–322 BC), who

* R. Markus, *The End of Ancient Christianity*, CUP, Cambridge, 1990, p. 224, quoting A. MacIntyre, *Whose Justice? Whose Rationality?*, Duckworth, London, 1988, p. 362

had the greatest influence on the development of Western philosophy in the years AD 500–1000. Essential to the reasoning of both these men (and to most leaders of Greek thought in the pre-Christian era) was a divide between matter and spirit. They embraced monotheism because the bevy of Olympian deities worshipped by their contemporaries was not capable of providing answers to their questions about the nature and purpose of the cosmos. They did not believe in a creator god because it seemed manifest to them that the material universe was eternal.

Plato, however, insisted that life must have *meaning*. His observation of the way everything functions led him to insist on the existence of *final causes*. There must be some rational entity outside material reality who fashioned crude matter into the myriad forms available to human senses and who produced each form for its own purpose. It is not difficult to see why medieval teachers latched onto this 'argument from design' as supporting evidence for the existence of the Christian God. But for Plato this being was remote and ultimately unknowable. Building on the dualism between pure essence and impure matter, the philosopher insisted that this same dualism was fundamental to the human condition. Man possessed a 'rational' soul within an imperfect and corruptible body. His felicity lay in curbing, as far as possible, his natural passions, so that his higher self could achieve intellectual enlightenment. By this self-control, not only would he lead the virtuous life, he would also be able to promote justice, peace and harmony in his society. Because the enlightened soul seeks greater unity with the divine, it seeks greater understanding of the natural world, for it is by investigating how things are that the philosopher is led to consider why things are – i.e. final causes. Abstract science had an important part to play in this process. For example, Plato projected the opinion that 'geometry is knowledge of the eternally existent'.

Aristotle is considered by many to be the greatest polymath the world has ever seen. He studied physics – by which he meant all aspects of the natural world – what seventeenth-century thinkers would call 'natural philosophy' and what would later become divided into the various 'natural sciences'. From there he moved into the realm of metaphysics (usually, though not strictly accurately, defined as 'higher physics'), which covers the implications of scientific discovery for abstract issues such as space, time and human nature.

His approach involved a combination of empiricism and speculation and there was no sense in which the latter was totally dependent on the former. He undertook a prodigious amount of hands-on research. He dissected numerous animals, examined the hive life of bees and calculated the movements of heavenly bodies. But his theorising was not always based on careful observation. He followed Hippocrates in asserting that the functioning of the human body was governed by the interaction of four 'humours' – blood, phlegm, yellow bile and black bile. He also stated that the speed at which objects fall to Earth is directly proportionate to their weight (something that simple experimentation would have disproved).

Such discoveries and reflections were among the intellectual baggage handed on to medieval scholars from their ancient forebears. Generations of monastic teachers and students treated this legacy with reverence and brought classical learning to bear on their theology. But did Greek thought provide the essential motivational force in the development of European scientific culture?

> Rapid intellectual and material progress began as soon as Europeans escaped from the stultifying grip of Roman repression and mistaken Greek idealism.*

That statement by Professor Stark in his thought-provoking study, *The Victory of Reason*, challenges any simplistic concept of the relationship between the classical and early medieval worlds that regards the

* R. Stark, *The Victory of Reason*, Random House, New York, 2005, p. 32

former as a high peak of civilisation and the latter as a descent into barbarism. Viewed in a wider historical perspective, the Roman Empire was just one of a number of aggressive ancient states based on slave labour and military might. Its pagan rulers claimed semi-divine authority. The minority of free subjects only possessed identity insofar as they were *citizens,* members of a corporate entity. Even during the period of the Republic (509–27 BC), political influence was only wielded by a small elite within this minority.

Because successive rulers had at their disposal a vast reservoir of slave labour as well as the goods brought to the capital and to regional centres by farmers and merchants, they were able to create an impressive lifestyle for their wealthy elite, making possible the enjoyment of art and literature and leaving to posterity an impressive heritage of private and public buildings and monuments. None of that should conceal the fact that it was a brutal regime, dependent on a well-trained army and sustained by conquest that ultimately collapsed because of its inherent structural weaknesses. But, as often observed, 'Conquered Greece conquered Rome'. This Mediterranean superpower absorbed much of Greek civilisation and became its conduit to the ages that followed. Senators employed Greek teachers for their children and scribes translated Greek writings into Latin. To be sophisticated in Roman society was to 'think Greek' and much of this culture survived when the power of the empire crumbled. In later centuries, Hellenistic philosophy entered the mainstream of European scholarship via Muslim and Jewish thinkers (see p. 12). Meanwhile, Europe's monasteries preserved and assimilated the incomplete elements of Greek philosophy that they could access.

Within their walls the speculations and scientific observations of the Greeks met with a third and more powerful strand of knowledge – revelation. Christian scholars applauded the endeavours of the old philosophers who had built stairways reaching into the clouds of the divine unknown but their Gospel spoke of the knowable God whose sunlight shafts had broken through from the clear, blue heaven. This it was that provided the motive force for medieval scientific thinking. In the closing years of the eighth century, the Northumbrian monk Alcuin travelled to Aachen to become principal of the court school set up by Charlemagne, the great emperor of the Franks. Among the correspondence of king and teacher that has survived is this observation by Alcuin of the intellectual basis of government in a Christian state:

> Many follow your well-known interest that a new Athens is being created in France, indeed a far finer one. For that which is ennobled by the teaching of our lord Christ surpasses all academic education; that which had only Plato's teaching owed its reputation to the seven arts, while ours is enriched by the seven-fold Spirit and so excels all earthly wisdom.*

Yet there remains a paradox at the heart of early medieval scientific endeavour. On the one hand, Christian theology turned the key and admitted the scientific quest into an intellectual mansion with numerous (potentially innumerable) rooms. Because the Creator had chosen to reveal himself it was possible for man to explore his works, to marvel at them and to render to him due worship. Not only was it possible; it was imperative for him to do so.

> Give thanks to the Lord because he is good;
> His love is eternal,
> Who can tell all the great things he has done?
> Who can praise him enough?
> Happy are those who obey his commands,
> Who always do what is right.
>
> 'Psalm 106', 1:3

* S. Allott, *Alcuin of York: His Life and Letters*, William Sessions Limited, York, 1974, p. 93

So ran one of the *Psalms* that were part of the daily ritual in abbeys and cathedrals throughout Western Christendom. Knowledge, worship and ethics were inseparable. Thus, for example, Bede, the seventh-century monk of Jarrow, is best known for his *Ecclesiastical History* but his writing embraced several disciplines and included *On the Nature of Things* and *On the Reckoning of Time*. Adjusting the calendar by reference to the relationship of Earth and Sun was important because it affected the calculation of the correct date of Easter. Bede was bringing scientific calculation into line with a coherent Christian theology – and making no distinction between the two. Much the same could be said of the scholars whose mathematical studies helped them to develop musical notation, a stepping stone from Gregorian chant to more complex polyphony.

Yet, on the other hand, while ecclesiastical librarians and teachers preserved the books of ancient wisdom, often keeping them safe while war and turmoil raged beyond the monastery walls, they regarded them with a veneration that discouraged debate. Not only was it regarded as heresy to dispute traditional interpretations of Holy Scripture, it was considered the next worst thing to heresy to disagree with the Greek pioneers of philosophy and science. Thus, although Bede demonstrated inaccuracies in the Julian calendar, these inaccuracies were not acknowledged and addressed until the old system was replaced by the Gregorian calendar almost eight centuries later.

It was in the 'practical sciences', where abstract disciplines merged into what we might call technology, that men had more freedom to discover new things and make use of them. Nowhere was the appliance of science more clearly apparent than in the development of church-building. It has often been observed that Gothic cathedrals and churches were scholasticism in stone (scholasticism being the intricate and demanding higher education system that had evolved by the eleventh century).

> Like the numerous all-embracing encyclopaedic constructions of 'high' scholasticism . . . the Gothic cathedral embodied the entire system of Christian knowledge and expressed the 'visible logic' of the cosmos.*

In the construction of these remarkable buildings, abstract mathematics came together with understanding of the qualities of stone and seasoned timber to produce intricate miracles of design for the glory of the Christian God.

Meanwhile, the Church had bestowed its greatest gift on European culture – universities: the new centres of learning which, by the mid-twelfth century, had begun to replace monastic and cathedral schools. Bologna, Paris, Oxford and other centres became the predominant purveyors of intellectual excellence. The courses they offered harked back to Plato's definition of the liberal arts and there was now a deepened interest in Aristotelian logic fuelled by rediscovered works preserved in Islamic libraries (see below). The student began his studies with the trivium – grammar, logic and rhetoric – then moved on to the quadrivium – arithmetic, geometry, music and astronomy. Having laid this foundation, the brightest scholars were ready to work for degrees in law, medicine or theology (the 'queen of the sciences'). Teaching took the form of two activities – the lecture and the disputation. Students were instructed in the wisdom of the past and the great doctors of the Church who had added to this depository through their own commentaries on the Bible and the Church Fathers. They were also trained in the art of dialectic – how to argue about the application of inherited truth to contemporary issues and problems.

What has been called the twelfth-century renaissance was just one of a number of periods of high culture that punctuated the

* A. J. Gurevich, *Categories of Medieval Culture*, Routledge, Oxford, 1985, p. 89

Middle Ages. The love of learning fostered over the centuries in European schools, monasteries and the first universities produced a growing class of 'professional' scholars. The threadbare academic was a familiar figure long before Chaucer immortalised the Clerk of Oxford in his *Canterbury Tales*:

> He would rather have at his bed's head
> Twenty books bound in black and red
> Of Aristotle and his philosophy
> Than rich apparel, fiddle or gay psaltery . . .
> Every penny he could wheedle out of friends
> Was spent on books and learning . . .
>
> Geoffrey Chaucer, 'The General Prologue', *The Canterbury Tales*, 1478

What the poet presents to us is the representative of a type; the member of an easily recognised class of unbeneficed clergy: the wandering scholars (*clerici vagantes*), who were devoted to learning and travelled long distances to track down rare manuscripts or sit at the feet of famous teachers. At times of comparative international peace there were hundreds of these peripatetic, knowledge-hungry men pilgrimaging freely along the network of roads that connected monasteries and other centres of learning. Thus it was that Western Christians made closer contact with Eastern Christians and also with their Muslim counterparts. In the libraries of the Byzantine Church and in Arabic schools in the Levant, Spain and Sicily they discovered hitherto unfamiliar ancient Greek texts and the commentaries of Muslim and Jewish scholars. These included the geographical and cosmographical works of the second-century Greco-Egyptian writer, Ptolemy, who, as well as conceiving a world map accepted as authoritative for a thousand years, made accurate observations of stellar movements, thus enabling horoscopes to be cast more accurately.

At this point some mention must be made of astronomy/astrology (the terms were interchangeable in pre-Renaissance Europe) which enjoyed a sudden flowering in the twelfth century. Study of the heavens had long been important to Christian monks, who had to read the night sky to know the right time for their offices during the hours of darkness and, as we have seen, scholars like Bede were concerned about calculating the appropriate timing of major festivals. But this area of research had other implications. Greek philosophy had imbued such studies with pseudo-magical content. To them it seemed self-evident that the movement of the 'stars in their courses' was directly related to the conduct of human affairs.

Early medieval scholars were ambivalent about astrology. Augustine rejected the idea that our fate is written in the stars because that would make a nonsense of free will. However, Boethius, who wrote his *Consolations of Philosophy* around AD 520, argued that, while human will is free, human nature is subject to the same immutability of Providence that governs the heavenly spheres. Even the Bible seemed to support the connection, for was not the birth of Christ heralded by the star that drew eastern astrologers to Bethlehem? Cures offered by monastic and other physicians were often believed to rely for their efficacy on the co-operation of the stars. For example, a tenth-century herbal remedy prescribed cannabis for the treatment of epilepsy and ordained that it must be ingested while the moon was in the sign of Virgo. The casting of horoscopes was based on the same principle, namely that there were propitious times for the undertaking of certain activities. Just as the movement of the Sun around the Earth determined the right seasons for planting and harvesting, so in all other human endeavours it was wise to fall into line with the divine design as mapped out in the heavens. Since most people – especially kings contemplating war and merchants investing money in trading ventures

– aspired to see into the future, the skilled astrologer was sitting on a potential gold mine.

It is not surprising to discover that one area of scientific exploration that was not the sole preserve of monastic practitioners was medicine. Disease and disability were rife and the ranks of those who offered treatment to the afflicted ranged from the local wise woman, the village midwife and the fairground mountebank to serious students of the human body. Care for the sick was a traditional ingredient of the cloistered life and, over the centuries, infirmarians experimented with herbal remedies and minor surgical practices. However, medicine as an academic discipline was fostered in lay schools, few of which seem to have been connected to worship centres. That of Salerno is the earliest of which we have knowledge. It came into being in the ninth century. Here the physician Gariopontus (died *c.*1050) wrote *Passionarius*, the first manual of known diseases and their treatments. What is more remarkable for the times is that in this town there also lived and worked a group of women – the 'Ladies of Salerno'. These, if the legends may be believed, were medical practitioners who plied their skills alongside men and even contributed to the literature of medical science.

Medicine, like law, was a very popular profession because it was very lucrative. That is not to say that many of the doctors did not apply themselves assiduously to the mastery of their craft. Yet few real advances were made in the Middle Ages, again because students were restricted by slavish veneration of the classical pioneers. The second-century physician, Galen, through his many writings, dominated medical science for one and a half millennia. He was as dogmatic as he was detailed. Like the standard works of scholastic theology, while providing an educational basis, his work also inhibited independent thought. Galen had asserted the importance of the study of anatomy. That was a significant breakthrough but, as in other areas of intellectual enquiry, speculation stood alongside empirical observation. Since experimentation on human corpses was forbidden, Galen based his deductions on the dissection of apes and pigs, asserting that their anatomies were similar to those of human beings.

It was through the writings and experiments of Arabic scholars that alchemy came to be established as a science in the West. Here we are entering the realm of 'magic'. The (disputed) origin of the word exists somewhere in the activities of early Persian *magi* and embraces both knowledge and power. The wise man was the specialist who not only understood the secrets of nature, but could manipulate them to his advantage. According to the hermetic tradition, deep-rooted in Egyptian, Greek, Jewish and early Christian thought, the origins of all knowledge were to be found in writings supposedly set down by the mystical figure of Hermes Trismegistus in the very dawn of time, when God made available to man *prisca theologia*, the foundation of all religions and philosophies. Those initiated into the threefold arts of theurgy (divine working), astrology and alchemy were possessed of awesome power, including divination, the ability to conjure spirits, the authority to cast spells and the means of transforming matter.

One example must suffice to personalise the spread of arcane speculation among the intellectual and political elite of medieval Europe in the early part of the last millennium. The favourite philosopher/physician/astrologer at the cultured court of the thirteenth-century Holy Roman Emperor, Frederick II, was known because of his country of origin as Michael Scot (1175–1232). This *clericus vagans* had travelled widely before joining the imperial retinue in Sicily but seems to have spent most of his time immersing himself in the hermetic teaching of Arabic, Jewish and Christian scholars in Spain, and reaching back to the ancient wisdom that predated all religions. Through such ecumenical groups, Greek and other early writings, hitherto unknown, reached the centres of European culture.

Among the twelfth-century translator-philosophers working in this intellectual forge were the Muslim, Averröes (1126–1198), and the Jew, Maimonides (1136–1204), both of whom laboured to reconcile classical wisdom with their respective religious traditions. The doctors of the Church similarly strove for a consensus between Aristotle and the Christian scriptures. The task was intricate and laborious. For example, how could a religious orthodoxy that insisted on a god who had created the universe be reconciled with the 'infallible' Aristotle, who had taught that matter is eternal?

But, to return to Michael Scot, thanks to his writings we have a record of the sort of problems exercising the minds of intelligent thirteenth-century enquirers. Frederick II posed to his court philosopher a number of questions, such as, how high is heaven in relation to Earth? Where is hell? Where is purgatory? There was nothing naïve or foolish about these questions. Given the accepted cosmography of the time, which placed Earth at the centre of a series of extrinsic circles reaching to the stars and beyond, it was reasonable for the emperor to expect his philosophers to provide him with detailed knowledge of the divine map.

By the late Middle Ages two contrary forces were at work in the intellectual life of Europe. On the one hand the Church continued to develop educational institutions and encourage rational inquiry. On the other it inhibited any developments that might be regarded as critical of received wisdom. Inevitably, these two impulses eventually clashed – violently. In 1022 a group of heretics were burned to death at Orleans. This was the first such execution to be sanctioned in more than 640 years but, over the next few centuries, heresy trials became increasingly frequent. Earnest men and women rejected aspects of official doctrine and some were prepared to die for their alternative faith. Few of them rejected the Christian basics. The majority were people who, for one reason or another, could not square the behaviour and teaching of their priests with what their intellects told them about God, nature and morality. One aspect of rebellion was anti-clericalism: critics of the ecclesiastical hierarchy were angry that clergy who wielded such power over them – in this world and the next – did not practise the holiness they advocated. Allied to this was scepticism about aspects of doctrine and custom. The wealth accumulated by the Church from clerical fees, the business of pilgrimage and ecclesiastical taxes began to smell of exploitation. It was but a short step to the questioning of some aspects of traditional teaching. Free-thinkers suggested that the Church had wandered into error and betrayed the Gospel entrusted to its care. Education was spreading and ceasing to be an exclusively clerical preserve. The development of legal and medical studies, as well as the growth of international trade, fostered new ways of thinking. Intelligent laymen (and, indeed, some clergy) read the Bible for themselves (in Latin originally, but increasingly in European vernaculars as portions of Scripture were translated) and decided that errors had resulted from the scholastic process of studying what generations of revered doctors of the Church said about the Bible, rather than getting to grips with the foundation document itself.

Ironically, it was the phenomenon of heretics coming to the thousand-year-old text with a new curiosity and excitement that alarmed the upholders of traditional interpretation.

> The secret mysteries of the faith ought not . . . to be explained to all men in all places . . . For such is the depth of divine Scripture that not only the simple and illiterate but also the prudent and learned are not fully sufficient to try to understand it.*

So declared Pope Innocent III in 1215 and behind that directive lay a fear that honest enquiry into the workings of God in his Creation

* Cf. M. Lambert, *Medieval Heresy*, Wiley-Blackwell, Oxford, 1992, p. 73

might weaken the Church's control of intellectual endeavour. Unlike the very pioneers of Christian thought whose reputation he sought to protect, Innocent regarded ancient wisdom not as a spur, encouraging the intellect to gallop forward, but as a bridle to hold it in check.

But no pope could fasten a padlock on all enquiring minds. Within ten years the greatest philosophical theologian of the Middle Ages was born. Thomas Aquinas was a brilliant dialectician who entered the debate on the relationship between Aristotelianism and Christian belief. He made his intentions quite clear: 'The study of philosophy is not done in order to know what men have thought, but rather to know how truth herself stands'*. His approach was essentially empirical: the search for truth must start from what *is*; what can be observed. That divides into two categories: the natural order perceived by the physical senses; and supernatural reality, which comes from divine revelation. Enlightenment derives from the application of reason – to *both* sources of knowledge. Since God is the originator of both, they cannot be in conflict. Aquinas produced a prodigious amount of written argument embracing a wide range of topics including theology, philosophy and ethics but his main concern (as developed in his hugely influential *Summa Theologiae*) was to employ logic in defence of Christian belief.

Other thinkers of the time were more involved in what we regard as 'experimental science'. Roger Bacon, a close contemporary of Aquinas, has acquired a spurious reputation as a 'real scientist ahead of his time'. In fact, he was very much a scholar *of* his time, well versed in languages, ancient and modern, and an eager student of Muslim and Jewish thinkers. His passions embraced optics, mathematics and astronomy, as well as theology. He fought against the

* M. D. Chenu, *Toward Understanding Saint Thomas*, Regency Publishing, Washington, 1964, p. 28

closed-mindedness of those who paid overmuch regard to the revered doctors of the Church instead of personal experience. Among his oft-repeated aphorisms is his statement about fire: someone who had never seen it might be convinced by reasoned argument that it burns but he would not *know* this until he thrust his hand into the flames. This does not mean that he was a forerunner of the post-Enlightenment materialist who could only trust what he could see, hear, taste, smell or touch. For Bacon there were two kinds of experience – that gained from the physical senses and that embraced by internal experience acquired by mystical communion with God. It is significant that Aquinas, Bacon and other like-minded adventurous thinkers found themselves in trouble with their ecclesiastical superiors, not because of their 'dabbling' in physical science, but because they were religious reformers challenging Church leaders to be more effective. Bacon, for example, urged his contemporaries to study the Bible in its original languages instead of relying on centuries of hackneyed interpretation.

We have, at last, arrived at the threshold of 'modern history' and the subject of our study but this preamble has been, I believe, necessary in order to make clear the well-established pattern of intellectual development in the West. As Professor Stark has said, 'The path to modern times did not suddenly open during the Renaissance any more than it sprang from the forehead of Zeus.'* Our civilisation evolved and our response to our environment was just one part of that process. Pioneer thinkers accumulated a growing body of *scientia*. But that was far from being the only gain made during these centuries. Philosophy, art, poetry and music helped to satisfy people's emotional needs and religion was the pack horse on whose broad back they – and scientific enquiry – were carried. Although they were not always in accord, science and superstition

* *Op. cit*, p. 68

travelled together during the centuries between the fall of Rome and the Renaissance. They continued to do so for a long while afterwards.

Nor was it only the rich and sophisticated who thought in these material terms. There was scarcely a church in Europe that did not feature as a major part of its décor a painted or sculpted 'doom'. This pictured Christ, the Judge, seated on a rainbow and consigning the blessed to a paradisal garden and the damned to a dark, fire-emitting cavern from which demons emerged to drag them to eternal torment. Such images cohabited, in the minds of ordinary people, with ancient folklore and its tales of wood-sprites and wizards, elves and enchantments, potions and prophesies. In a world where life-expectancy was short and disease, famine and war were frequent visitors, most people did not discriminate nicely between those they turned to for supernatural aid. The fundamental questions in most minds were 'How can I survive in this world and ensure safe passage to the world to come?' and 'What must I do to be saved?'

The church had developed its own magic to provide answers to these questions. Whatever philosophers and theologians might debate in the new universities, practical, everyday religion offered – nay, demanded acceptance of – a variety of spells, charms and rituals designed to improve behaviour in this world and ensure blessedness in the next. Holiness was attached to certain places and objects. The devout (or desperate) seeking cure for their ailments might travel to a distant shrine to touch a reliquary holding a fragment of some long-dead saint. Above all, they needed reassurance about the eternal destiny of themselves and their loved ones.

This affected burial rites. There was competition for acquiring 'blessed' grave sites in churches (the closer to the altar, the better). Some prominent members of society were buried in monastic habits. Official doctrine declared that after death the heaven-bound soul had to pass through a period of purification before being admitted to the presence of a holy God. According to the doctrine of purgatory evolved by the thirteenth century, the duration of this experience could be shortened by the prayers of the saints and of living clergy. Thus, the wealthy might buy masses to be performed regularly by priests. For example, Henry VII of England left funds for ten thousand masses to be performed immediately after his death in 1509 and another fortune for still more to be said in perpetuity. Poorer folk were left with little to rely on but their own virtue.

The mass was, indeed, powerful magic. Performed by a caste of priestly alchemists, it constituted the transmutation of bread and wine into the body and blood of Christ. The awkward fact that the elements looked, smelled and tasted exactly the same after the celebrant had muttered his spell over them was disposed of by twelfth-century scholastic theologians with an argument based on the Aristotelian distinction between the *substance* of an object (i.e. its essence) and its *accidents* (its outward form). Thus, it was averred, consecration transformed the substance of bread and wine (their 'breadness' and 'wineness') while leaving the accidents unchanged.

CHAPTER TWO

A magical world

In 1500, the average life expectancy in Europe was thirty-three to forty, figures somewhat skewed by the high rate of child mortality, frequent epidemics and by social distinctions.

Some 30 per cent of infants died before the age of five. The majority of people were not well nourished and lived in what we would now consider slum conditions, in which disease and fire hazard lessened the likelihood of survival beyond forty. By contrast, the very different lifestyles of noble and mercantile families and of the cloistered inhabitants of monasteries and nunneries was reflected in their comparative longevity.

Death was an almost daily occurrence in most communities, and if people were tempted to close their eyes to it, they were reminded of its reality every time they entered a church or passed by a charnel house with vivid images of the skeletal 'Dance of Death' or the lurid 'Last Judgement' representations of Christ consigning the departed to heaven or hell. There can have been very few people who, when confronted by illness, injury or some other life-threatening circumstance, did not reflect on the question, 'What must I do to be saved?' The answers they found lay in the spiritual realm.

The Church provided access to that. Through its rituals it claimed to connect mortals with 'the saints in glory'. Every medieval will began with the bequest of the testator's soul to God with the plea that the Virgin Mary and the denizens of heaven would pray for it. As we have seen, those who could afford to do so left money for the performance of masses, so that on Earth priests could add their voices to those of the saints. It was in specific answer to the question, 'What must I do to be saved?' that the doctrine of purgatory had been further developed. Such beliefs were not unique to Christendom. Prayers and incantations for the dead pre-dated Jesus and were employed by the followers of other religions. They represent a very natural human reaction to the mysteries of life and death. But the medieval Western Church systematised purgatory, combining with classical cosmography the spiritual problems of the sinful soul's achievement of perfection and acceptance into heaven.

> Meanwhile we had come up to the mountain's flank
> There at its foot we found the rock so sheer,
> Vainly would legs be limber on that bank.*

In the *Divine Comedy* Dante (1265–1321) pictured purgatory as a mountain ('the Mount where Justice probeth us'). Its painful ascent represented the process by which the Christian soul could complete that process of sanctification that alone could fit it to enter the presence of God. Much influenced by Aristotelian concepts of vice and virtue, the Church taught that there were some sins that could not be fully atoned for in this life – hence the need for this intermediary stage. In Dante's cosmography, paradise could only be reached from the summit of Mount Purgatory. It would then pass through the nine concentric spheres representing the nearer heavenly bodies and the fixed stars and so reach the *Primum Mobile*, the First Cause.

Dante's masterpiece was poetic allegory but we should not think that it was totally distinct from official teaching about *spiritual* realities. The medieval mind (certainly the untrained medieval mind) made no distinction between the observable universe and the spiritual realm. For most people, the world beyond their own country (and, for some, even beyond their own market town) was a mystery. When travellers returned from distant lands with stories of monsters

* P. Milano (ed.), *The Portable Dante*, Penguin, London, 1977, p. 200

and men with faces in the middle of their chests they were readily believed, for no one had any knowledge to the contrary. Whether or not more sophisticated minds embraced the idea of purgatory as a *place*, such beliefs enabled people to engage their imagination and to relate life here to life hereafter.

To be sure of a short and successful passage through purgatory people had to avail themselves of the means of grace provided by the Church. These did not only apply to blessings in the world to come; medieval man was interested in making life as bearable as possible in the here-and-now. He believed that religious rituals enabled him to harness spiritual powers to help him deal with the crises of daily life. This was where relics came in. These were bodily fragments of dead saints or items associated with them, housed in churches and monasteries where the devout could gaze upon them and even touch them. Pilgrimage to such shrines fulfilled two purposes. Saints were believed to have authority over certain maladies and would respond to the prayers of the faithful by granting relief. For example, St Roch was good for plague, St Quirinus for deafness, St Apollonia for toothache and so on.

Visits to shrines and offerings made there also counted as pious deeds that merited reward. From this sprang the custom of granting 'indulgences' which, by the fourteenth century, normally took the form of certificates granting relief from specified periods of time in purgatory. Talismans, pilgrim badges and other items of holy significance were worn as charms against evil. Intelligent people, not unnaturally, wanted evidence that such religious paraphernalia actually worked. By way of response, they were offered miracles.

Miracles were the stock-in-trade of the Church. They were the 'proofs' of Christian truth and encouragements to faith. Central to the life of every believer were the sacraments and, particularly, the mass. Every priest was a routine wonder-worker. By intoning a few words over bread and wine, he was able to transform their nature into flesh and blood. As we have seen, over the centuries the Church had refined its explanation of what actually happened at the altar but sophisticated analysis of the 'accidents' of bread and wine and their 'substance' were lost on most worshippers. It was not their role to understand; they were simply present as spectators of priestly magic. This routine miracle was accepted and so it was not difficult for people to accept accounts of more dramatic happenings. Wondrous anecdotes featured prominently in sermons and were represented in stained glass and paint on the walls of churches. One of the early books in English published by William Caxton, who was thought to have introduced the printing press to England, was a 1483 translation of *The Golden Legend*, a lengthy collection of anecdotes about saints and martyrs, written two centuries earlier by Jacobus de Voragine, Archbishop of Genoa. A staple of preachers throughout Europe, the book became a runaway bestseller and went through nine editions in its first half-century. The kind of eye-widening stories about the miraculous power of holy relics that enthralled readers is illustrated by the tale of the pall of the martyr, St Agatha.

> One year from the day of Agatha's birth into the new life of heaven, the mountain that looms over Catania erupted and spewed a river of fire and molten rock down toward the city. Then crowds of pagans fled from the mountain to the saint's tomb, snatched up the pall that covered it and hung it up in the path of the fire and . . . the stream of lava halted and did not advance a foot farther.*

Such wondrous tales may have impressed most medieval men and women but they also created problems for the ecclesiastical

* W. G. Ryan (trs.), *The Golden Legend: Readings on the Saints*, Princeton University Press, New Jersey, 1993, I, p. 40

hierarchy. Theologians found themselves on the horns of a dilemma: how could they claim that miracles, performed by God in response to the petitions of the faithful, were different from signs and wonders manifested by magicians? If it was laudable to pray to the Virgin for blessing upon a commercial venture, why was it reprehensible to obtain a love potion from an apothecary or recite magical incantations to discover the identity of a thief? If both priests and magi presented themselves as conduits of supernatural power, what was the difference between them?

The stock answers to these questions were presented in such books as the *Livre de Tresor*, by the Florentine scholar Brunetto Latini (d. 1294). This prominent and talented notary served his native city in various capacities (despite having to spend several years in exile as a result of political faction fighting) but he is particularly famous for being the much-loved guardian of Dante Alighieri following the death of the poet's father.

The *Livre de Tresor*, written during Latini's sojourn in France, was the first European encyclopaedia. It described occult practices as originating from Zoroaster, the Persian philosopher/mystic/magician. He it was who 'discovered the magic art of spells and other wicked words and wicked things . . . during the first two ages of the era that finished in the time of Abraham'*.

What distinguished pagan magic from Christian magic was the identity of the beings whose aid was being sought. Exegetes drew on references in the Bible and ancient Jewish texts to show that magicians obtained their powers from the conjuration of demons and that their motives were the pursuit of wealth and personal adulation. Jannes and Jambres were villains who, according to the Talmud, set themselves up in opposition to Moses. Simon Magus was a Samaritan magician who tried to buy the secret of divine power displayed by the

* Bruno Latini, *Livre de Tresor*, Gondoliere, Venice, 1839, p. 33

apostle Philip. Christian tradition insisted that Simon had come to a sticky end by practising levitation and being returned precipitately to Earth by the prayers of Saints Peter and Paul.

In the medieval cosmos the spiritual realm overlaid the material. Everyone knew that and everyone turned to the experts – priests, magi, astrologers, wise men and wise women – to help in their journey from cradle to grave – and beyond. From the highest to the lowest, people trusted to Christian magic or pagan magic – or both. Kings had particular need of guidance and protection from the spirit world in managing their affairs. They employed not only sage churchmen as councillors but also astrologers to advise them on propitious dates for making war, agreeing treaties and receiving diplomatic missions, etc. and wise men (from which comes 'wise-ards' = wizards) to conjure the power of the spirit world on their behalf.

The legends of King Arthur and Merlin, who were considered to be fully historical characters, were popular in English royal court circles and indicate how natural such relationships were in the political centres of the medieval world. Pagan magicians were always something of an embarrassment to Church leaders but they were not officially outlawed. Not until 1564 were collections of Merlin's prophecies placed by the Vatican on the list of banned books, the *Index Librorum Prohibitorum.*

Various ways were found to Christianise the Merlin corpus. One tradition claimed that the wizard had come under the influence of a holy priest who had sanctified his occult gifts. It would be difficult to overemphasise the influence of the Merlin tradition throughout the period we are considering. We shall return to this later when we think about prophecy, for an ever-growing corpus of predictions was attached to this Arthurian magus. As late as 1641, Thomas Heywood, in his *Life of Merlin,* described how God, who is free to reveal his purposes in any way he chooses, had selected the pagan seer to foretell the things that must come to pass.

When we explore the world of 'popular magic' we discover a range of beliefs and practices as diverse as they were fantastical.

> In the olden days of King Arthur,
> Of which the Britons speak great honour,
> This land was all filled with fairies.
> The elf-queen with her jolly company
> Danced full oft in many a green meadow.
> This was the old opinion as I read.
> I speak of many hundred years ago.
> But now can no man see elves any more.
>
> Geoffrey Chaucer,
> *Canterbury Tales*

Chaucer's Wife of Bath was happy to consign belief in aerial sprites to a long bygone age – not because such creatures never existed, but because the prayers of holy men had banished them from most of their old haunts. Yet, two hundred years later, Shakespeare could still people the Forest of Arden with Oberon, Titania and their fairy courtiers; frighten Falstaff with tales of imps and elves in Windsor Great Park; and present to his audience the awesome magus, Prospero, who had spirits at his command. That is not to say that, by the turn of the seventeenth century, Robin Goodfellow – Puck – and his ilk still captured the imagination of all English people but that fairy stories had not yet been confined to children's literature. In fact, they would be an unconscionable time dying.

It is significant that medieval Christian teaching had to find a place for fairies, elves, goblins and other mysterious 'little people'. According to one tradition they were fallen angels, a class of spiritual beings somewhere between angels and demons. These characters from pagan myth, whose activities were passed on orally in stories down the generations, were so firmly entrenched in folklore that the

Church could not ignore them. Though generally regarded as benign, the various orders of sprites were believed to have real powers and it was unwise to cross them. Generations of country wives left out food for the fairies overnight so that the invisible visitors would not inflict disease on their livestock or abduct babies from their cradles. Children with learning difficulties were thought by many to have been enchanted by the 'little people' or were, as we might say today, 'away with the fairies'.

Belief in such creatures was not confined to illiterate country folk. One of the most notorious witches of the fifteenth century, Margery Jourdemayne, claimed to be able to conjure 'fiends and fairies' and make them do her bidding. This wise woman, known as the Witch of Eye, came to a sticky end when she became involved with the affairs of the royal court. Though the wife of a mere cowherd, she had long enjoyed a reputation as a purveyor of potions, spells and charms to 'the quality'. Among her patrons was Eleanor, Duchess of Gloucester, whose husband, Duke Humphrey, was uncle to King Henry VI and, more importantly, heir presumptive to the throne.

Eleanor was accused of hatching a plot to advance her husband's promotion by eliminating the king. Her accomplices were her clerk, Roger Bolingbroke, an Oxford scholar – renowned as a 'great and cunning man in astronomy' – Thomas Southwell, a physician and canon of Westminster, and John Hume, the duke's chaplain. Bolingbroke confessed that he had produced a horoscope predicting the king's imminent death and revealed that Margery had made wax effigies of Henry with the intention of bringing about the fulfilment of the prophecy. After the 'guilty' verdicts had been delivered, Eleanor was divorced and spent the rest of her life confined in various royal castles. Bolingbroke was hanged, drawn and quartered. Margery was burned at the stake. The others died in prison.

Popular magic was a rainbow-hued phenomenon:

> You have heard of Mother Nottingham who, for her time, was prettily well-skilled in casting of waters and, after her, Mother Bomby; and then there is one Hatfield in Pepper Alley, he doth pretty well for a thing that's lost. There's another in Coleharbour that's skilled in the planets. Mother Sturton in Golden Lane is for fore-speaking; Mother Phillips, of the Bankside, for the weakness of the back and then there's a very reverend matron on Clerkenwell Green good at many things.
>
> Mistress Mary on the Bankside is for erecting a figure; and one (what do you call her?) in Westminster, that practiseth the book and the key and the sieve and the shears: and all do well according to their talent*.

That catalogue by the dramatist and essayist Thomas Heywood, writing in 1638, is testimony to the variety and durability of valued specialisms. Villagers and townspeople resorted to those of their neighbours who were thought to have certain skills. Herbalists, apothecaries, cunning men, wise women, diviners, prophets, casters of horoscopes, necromancers – all were thought to possess knowledge (and, therefore, power) of 'natural magic' and it was as much an everyday practice for our ancestors to seek help from such gifted practitioners as it is for people today to consult trained doctors, psychiatrists or veterinaries.

There were established practices for achieving certain ends. For example, the sieve and shears was an accepted method of unveiling a thief. The shears' blades were stuck into the rim of the sieve. Two people would lightly support the handles. The magus would then invoke the aid of Saints Peter and Paul. Each suspect in turn was named and the sieve would rotate when the culprit was mentioned. This was one of several quite routine rituals. Sometimes the

* T. Heywood, *The Wise-woman of Hogsdon*, (1638), Scholar's Choice, New York, 2015, III, p. i

officiating magus was the parish priest. When the criminal had been unmasked the constables would be quite ready to make an arrest.

This was just one aspect of natural magic which featured in the everyday life of all communities. It was a means of tapping into the unknown – the occult. An English Act of Parliament of 1542 indicates some of the more popular kinds of services offered by magicians to their potential customers and extends the death penalty to those using divination for the recovery of stolen property, treasure seeking or 'provoking any person to unlawful love'. Government concern at practices that were mischievous or downright satanic was not misplaced. Seven years later, William Wycherley, a London tailor on trial for conjuring spirits, told the court there were in England more than five hundred magicians adept at summoning demons. He described how some operated with crystals, mirrors, magic circles, consecrated swords and other paraphernalia to impress their clientele.

But this was not the only vibrant magic tradition in Renaissance Europe. The fate of the unfortunate Roger Bolingbroke indicates an intellectual stream that, at some points, flowed into the more naïve popular river. As the New Learning spread, the official attitude of the Church was to maintain a clear distinction between it and what went on in the market place, where simple people had their fortunes told or fell victim to the wiles of witches and wizards. 'Witchcraft is not taught in books, nor is it practised by the learned but by the altogether uneducated,'* as the *Malleus Maleficarum* asserted in 1486. This notorious book by two German Dominicans was a manual on how to identify, interrogate and punish satanic agencies. It was written at a time of growing concern at the spread of heresy and erroneous religious beliefs and practices.

The first objective of the authors was to scotch the belief of some scholars that witchcraft did not exist outside the imagination of

* M. Summers (trs.), *Malleus Maleficarum*, Pushkin Press, London, 1951, II. 1

those who practised it. They asserted that it was real and diabolical. The second objective was to assert that there was no connection between the sorcery and necromancy practised by the unlearned, on the one hand, and scholarly enquiry into the wondrous workings of the cosmos, on the other. The presupposition on which the book was based was that society was strictly stratified. God had set every person in his/her station. Social mobility, as well as being politically dangerous, was against the divine order. Witchcraft, like heresy, was confined to the lower orders and it was the responsibility of rulers, guided by their well-educated advisers, to stamp it out. The *Malleus* advocated the most extreme and, to the modern eyes, inhuman methods to eradicate this demonic intrusion into the world.

From the start the Christian hierarchy was uneasy about this draconian and uncompromising book. It was actually banned by the Vatican in 1490 and, though frequently printed in continental Europe, it did not become a persecutor's textbook until the seventeenth century. In asserting the distinction between the two kinds of magic – popular and academic – it bore little relation to reality.

Bolingbroke was by no means unique as an educated student of the occult becoming involved with popular sorcery. Intellectual magicians at this time were developing a close interest in the folkloric origins of the practices of witches and wizards. So far from rejecting all old knowledge, Renaissance scholars wanted to incorporate ancient wisdom into Christianity. The Renaissance produced a new breed of magi-men whose enquiring minds drove them to the edge of Christian orthodoxy and beyond. One such was Henry Cornelius Agrippa (1486–1535).

He first appears as a pre-Reformation reformer, an evangelical *avant la lettra*, one of a growing number of intellectuals like Desiderius Erasmus, Sebastian Brant, Jacques Lefèvre d'Etaples and John Colet who were critical of the state of the Church and wanted to bring it into line with biblical simplicity. Agrippa grew up in

Cologne but, like other truth-seekers, became an academic pilgrim, wandering from Germany to England, from England to Italy, from Italy to the free imperial city of Metz, and from there to Geneva, Flanders and France, before his death in Grenoble.

In his restless career he spent time as a soldier, student, physician, lawyer, theological lecturer, writer and adviser to princes. The influences on his thinking were as varied as the currently fashionable Neoplatonism (see below) and the new Biblicism of Martin Luther and other reformers. He was a part of 'that spiritual force which was breaking down the past and ushering in the future'.* Like Erasmus, whose humanist schema was a bringing together of classical studies and the Gospel in a *philosophia Christi*, Agrippa aimed to fuse evangelical theology with ancient strands of occult *scientia*.

Neoplatonism was the 'in' approach to universal knowledge popularised by the late fifteenth-century Florentine thinkers, Marsilio Ficino and Pico della Mirandola. What these philosophers were seeking was the *prisca theologia* – the origin of all religions and philosophies. The metaphysical system they evolved conceived of the universe as a living entity in which all elements were spiritually interconnected. This being so, the natural philosopher's role was to investigate and explain the workings of the cosmos, while the natural magician occupied himself in exploiting the latent power within the cosmos for the benefit of humankind. There were three categories of magical activity. Natural magic harnessed the forces present in the terrestrial sphere – plants, the elements of fire, air and water and the human physiognomy. Celestial magic concerned itself with the influence of the heavenly bodies. Ceremonial magic involved invoking the aid of spiritual beings.

* F. Yates, *The Occult Philosophy in the Elizabethan Age*, Routledge & K. Paul, Oxford, 1979, p. 41

Agrippa's synthesis was set forth in two major works. They reveal not only his philosophy, but also the intellectual struggle involved in framing it and defending it in the intellectual world of the early sixteenth century. *De Vanitate Scientiarum* (1533) was nothing less than a rejection of all human effort to comprehend the divine. Like the Old Testament wisdom writer of *Ecclesiastes*, Agrippa dismissed such labours as 'vanity'. He worked his way systematically through all the *scientiae* presented in the medieval scholastic curriculum and similarly scorned the new or revived disciplines of metaphysics, astrology, mathematics, alchemy, etc., etc. The only source of wisdom, he concluded, was the word of God. That would seem to place him squarely among the ranks of the Lutherans and, perhaps, of Erasmians. *De Vanitate* was in the vein of Erasmus' *Praise of Folly*, a 1511 exposé of the presumption of human endeavour to probe the mysteries of the cosmos. It struck a common chord with many readers who were suspicious of scholarly assertions. For while philosophers and magi claiming a deep understanding of nature and supernature were held in awe by some ordinary mortals, there were always sceptics who rejected what they regarded as intellectual pretension.

If the *De Vanitate* had been Agrippa's last word it would be easy to allot him his place among Renaissance thinkers but he went on to publish *De Occulta Philosophia*, in which he adopted a very different stance. It was one thing to sing with the chorus of humanist scoffers against the outworn dogmas of the Church and the inadequacies of newfangled philosophies but such negativism was not very helpful. It was incumbent on the responsible magus to come up with a positive schema that would enable man to discover true wisdom and his place in the divine ordering of the universe. This involved assessing hermeticism, cabala and other ancient intellectual disciplines in the light of Christian Scripture and direct, personal religious experience. Agrippa was trying to hold together everything he knew and felt and believed. *De Vanitate* had insisted that no one could enter heaven merely by *thinking* but, since God had given man the gift of logical thought, it was incumbent on him (or, at least, on those especially intellectually gifted) to make sense of the occult tradition.* The result was *De Occulta Philosophia*, which circulated in manuscript as early as 1510, though it made its first appearance in print only in 1533. In this book we are back in the metaphysics of Pico della Mirandola, the mathematical magic of Hebrew numbers and letters and, indeed, in the speculations of Hermes Trismegistus. Ptolomaic cosmography of circles within circles provided the pattern for Agrippa's understanding of the different layers of knowledge:

> There is a three-fold World, Elementary, Celestiall, and Intellectual, and every interior is governed by its superior, and receives the influence of the virtues thereof, so that the very original, and chief Worker of all doth by Angels, the Heavens, Stars, Elements, Animals, Plants, Metals and Stones convey from himself the virtues of his Omnipotency upon us, for whose service he made, and created all these things.

Agrippa insisted that this was not merely a subject for scholarly observation:

> It should be possible for us to ascend by the same degrees through each World, to the same very original World itself, the Maker of all things and first Cause, from whence all things are, and proceed; and also to enjoy not only these virtues,

* Cf. J. S. Mebane, *Renaissance Magic and the Return of the Golden Age*, University of Nebraska Press, Lincoln, 1992; C. G. Nauert, *Agrippa and the Crisis of Renaissance Thought*, University of Illinois Press, Illinois, 1966

> which are already in the more excellent kind of things, but also besides these, to draw new virtues from above.

In a word: 'magic'. Agrippa aimed to show not only that celestial and super-celestial bodies (angelic and demonic spirits) influence terrestrial life, but that they could be directed by mortals. In other words, Agrippa essayed to instruct his readers in the art of conjuration. His logical progression through the three spheres is simple and straightforward. The student of the elemental world learned about natural substances and the occult sympathies between them (the basis of medical practice). This led naturally to celestial magic; the influence exercised by the stars on the lower world. Finally, the adept would reach the super-celestial world in which ceremonial magic could summon the aid of angelic spirits and, ultimately, the Father of spirits.

Agrippa moved in the highest circles and numbered the Emperor Maximilian I and Francis I of France among his patrons. This, however, did not insulate him from criticism. Unsurprisingly, his teaching on the magical invocation of spirits aroused the interest of the Inquisition and he thought it wise to backtrack. In the published version of *De Occulta Philosophia*, he claimed that his writings had been misquoted and misunderstood and he put on record a solemn retraction. After mature reflection and deeper study, he declared his rejection of 'erroneous' statements and his conviction that

> Whosoever do not in the truth, nor in the power of God, but in the deceits of devils, according to the operation of wicked spirits presume to divine and prophesy and practising through magical vanities, exorcisms, incantations and other demoniacal works and deceits of idolatry, boasting of delusions, and phantasms, presently ceasing, brag that they can

> do miracles, I say all these shall with Jannes, and Jambres, and Simon Magus, be destined to the torments of eternal fire.

Agrippa's backtracking is testimony to the growing anxiety within the Church at the uncontrolled quest for knowledge (and, therefore, power) and where it might lead (as the *Malleus Maleficarum* indicated). Not all *scientia* was good – i.e. God-inspired. Some was generated by the devil and his cohorts. Enter Doctor Faustus.

The most remarkable fact about the Faust legend is that it *was* always a legend, rather than a collection of stories built up over a foundation of sensational fact. The anecdotes that began spreading through Europe in the early decades of the sixteenth century were embroideries on a narrative about a man (or possibly two men) whose actual identity was, at best, hazy. Even the name may be no more than a label derived from the Latin *faustus*, meaning 'fortunate'. The fact that the story of a scholar who dabbled fatally in arcane knowledge exercised such a powerful influence tells us a great deal about the corporate psychology of Renaissance civilization. People were ready for this cautionary tale. It fed their fears. It supported their suspicions. It satisfied their imagination. They had no difficulty in believing it. We may reasonably call it a fable whose time had come.

Johan Fust, the partner of printing pioneer Johannes Gutenberg, and Georg Faust, a late fifteenth-century necromancer, have both been suggested as the real-life foundation for this powerful myth. But it does not really matter who the original Faust was or whether there ever was an actual historical figure. What we need to focus on is the range of common attitudes towards the pursuit of knowledge.

Johannes Trithemius (1462–1516) was a Benedictine abbot, monastic reformer, scholar, historian and preacher. He was also a magician consulted by many students of the arcane (including Agrippa). It was he who, in a letter of 1507, provided the first description of Faust. This was not a flattering assessment. The man was, in Trithemius' estimation, 'a wandering vagrant, a driveller and a cheat', a dangerous fellow who taught things 'against the holy church'. The writer scoffed at Faust's claims to have mastered necromancy, astrology, chiromancy (palmistry), pyromancy (divination by fire) and aeromancy (divination by atmospheric conditions).* Was this genuine indignation directed at someone who had stepped beyond the bounds of Catholic orthodoxy or the jealousy of a rival practitioner?

Because the border between Christian orthodoxy and pagan practices was extremely porous it is difficult to assess where scholars drew the line in their love of mystique. Humanist studies probed well beyond the conventional bounds of knowledge and this placed them in a class of their own, a distinction the international clique of radical intellectuals enjoyed and deliberately fostered. In a sense they were all magi, delving into cosmic mysteries, sharers of ancient secrets. This lifted these Renaissance celebs above the world of ordinary mortals, winning them admiration bordering on worship (as well as material reward).

We can see how much they relished this in many of the artworks and writings that have survived. In-jokes and hidden messages were all part of Renaissance stock-in-trade. Concealed numbers and letters have been detected in Da Vinci's *Mona Lisa*. Holbein's *The Ambassadors* positively bristles with the artist's comments on contemporary events. Experts have detected cabbalistic signs in Michelangelo's Sistine Chapel frescos. Trithemius certainly loved playing with codes and performing verbal tricks. His major work,

* L. Ruickbie, *Faustus – The Life and Times of a Renaissance Magician*, The History Press, Stroud, 2009, pp. 39–40

Steganographia, is all about hidden messages. Ostensibly a study of magic and particularly the employment of angels to convey thoughts through the air, it was, for those who could decipher it, a treatise on secret writing. Yet we must be careful not to regard these creative minds as simply indulging in intellectual games. There were important and serious issues at stake. The Inquisition was very active and ready to pounce on anything that might be considered heretical. The Vatican was diligent in adding to its list of banned books. Some of Erasmus' works were among those that had the distinction of appearing on the *Index Librorum Prohibitorum* and, in 1609, *Steganographia* shared that honour.

It was generally assumed that not everything which might lie beyond the horizon of the known could be good. Agrippa was not the only scholar to study spiritual forces to determine whether or not magic could be divided into 'white' and 'black' categories. One could not be a true son of the Holy Church without acknowledging the power of holy relics and religious talismans. But Holy Scripture, as we have seen, recognised and denounced the conjuration of super-celestial forces to gain knowledge or power. King Saul in *I Samuel* 28 was condemned for calling on the Witch of Endor to summon up the ghost of the prophet. In *Acts*, chapter eight, Simon the Magus was rebuked for dabbling in magic arts and for presuming that he could obtain from the apostles the 'spells' that would bring spiritual enlightenment. Writers over the ensuing centuries were fascinated by this character who appeared briefly on the biblical stage. They created biographical details of his later life, one of which placed him in Rome using the Latin name 'Faustus'.

Medieval sermons were replete with stories of misguided folk who did deals with the devil. The thirteenth-century Cistercian prior, Caesarius of Heisterbach, told of two wandering miracle workers who attained great repute as a result of obtaining powers from the devil. They were only exposed when a holy man, posing as a satanist, tricked the devil into revealing how he exercised power over his agents.

In contrast with such cautionary tales, stories of demonic pacts in folklore were often presented as farce. Anecdotes were handed down through the generations of the devil being cheated of his prey by sharp-witted peasants. These provided one strand in what would eventually emerge in the traditional Punch and Judy show.

CHAPTER THREE

Re-evaluation

In 1451 King John II of Cyprus obtained from Pope Nicholas V a grant to sell indulgences, which were promises of reduction of time spent in purgatory in return for contributions to the Church, of which John took a cut. The king was in urgent need of funds to stave off the expansionist regime of the Ottoman Turks, a threat that became all the more menacing two years later when Sultan Mehmet II captured Constantinople and put an end to the Byzantine Empire. Needing to sell as large a number of these documents as possible, John despatched his agents to Mainz. Why Mainz? Because there a certain Johannes Gutenberg had established a workshop that could turn out printed sheets far faster and cheaper than anything the world had ever seen.

It is instructive to discover this direct link between the revolutionary development of the movable-type printing press and that pressing question which, as we have seen, occupied medieval minds: 'What must I do to be saved?' Gutenberg's invention, which has, with good reason, been claimed as the most important in the history of our civilisation, was a response to the growing demand for religious literature. For decades the monastic scriptoria had been hard-pressed to meet the need for standard Latin texts and devotional works. As a result, book production became an industry dominated by commercial scribes, unconnected to the religious houses, who helped to meet the rapidly increasing demand. The main contributor to that demand was the education sector. In Germany alone there was a fourfold increase in the number of students attending university in the second half of the century. This, in turn, was the result of an even larger number of boys receiving primary education. The single most popular book coming off the new presses was the *Ars Grammatica*, a tutorial on Latin grammar by the fourth-century pedagogue, Aelius Donatus.

Printing became the boom industry of the second half of the fifteenth century. Artisans rushed to cash in on the new technology and there was no lack of capitalists ready to back their enterprise. By 1500, every city and major town in Europe boasted at least one print shop and there were already in circulation a staggering nine million volumes. The more popular books carried woodcuts (replacing the illuminated illustrations of medieval manuscripts) but this did not mean that most readers could not cope with plain text. Statesman and councillor to Henry VIII, Thomas More, reckoned that three fifths of England's male population could read and, although this estimate is unreliable, it does signify a situation that contemporary observers found noteworthy. By More's time Europe had reached what Jacques Barzun called 'the age of indispensable literacy'.*

It was also the age of inescapable new knowledge. Maps based on Ptolemy's *Mappa Mundi* would no longer serve when mariners in the pay of Castile-Aragon and Portugal reached the coastline of the Americas and rounded Africa to establish direct contact with the Orient. The medieval basis of philosophical/theological knowledge also became insecure as refugees fleeing before the Ottoman horde brought with them classical documents rescued from Byzantine libraries – some hitherto unknown and others more accurate than most of those available in the West. Now the works of Plato and Aristotle could be seen in context as parts of a rich, long-lost heritage. Homer, Thucydides and Demosthenes were added to the literary relics of ancient civilisation, as were Latin poets, dramatists,

* J. Barzun, *From Dawn to Decadence* (1656), HarperCollins, London, 2000, p. 54

historians and natural scientists of the stamp of Virgil, Plautus, Tacitus and Pliny.

Add to such stimuli growing criticism of the status quo from observers as different as freethinking classical scholars and semi-educated heretics and you have a rare phenomenon: a self-conscious age aware of disturbing change. Just as academics were excited by the newly available Greek and Latin texts, so heretical fringe groups, such as Wycliffites in England and Hussites in Bohemia, like Cathars and Waldensians in previous centuries, drew inspiration from closely guarded vernacular fragments of Christian Scripture. As early as 1408, Church leaders in England had forbidden the translation of the Bible into English or any other language and, although this proscription was not echoed everywhere, authorities were on their guard against individuals or groups presuming to study the basic Christian text for themselves, without direction from their clergy.

Predictably, attempts to suppress free thought were as ineffective in the fifteenth as they had been in earlier centuries. They did not – immediately – provoke mass rebellion but they did encourage a mood of sceptical questioning and a tendency towards a new individualism. The two roaring bestsellers in Europe as the new century approached were books which, on the face of it, were very different. In 1486 the mystic Thomas á Kempis issued *The Imitation of Christ*, a call to personal holiness. 'What have we to do with the dry notions of logicians?' he demanded in a scarcely veiled reference to scholastic theology. 'He to whom the eternal Word speaketh is delivered from a world of unnecessary conceptions.'* Eight years later, a German scholar deplored the barbarity of his own age in comparison with the refined civilisations of Greece and Rome:

* Thomas á Kempis, *The Imitation of Christ*, Book 1, Chapter 3

> Knowledge of truth, prudence and just simplicity
> Hath us clean left for we set of them no store.
> Our faith is defiled love, goodness and pity.
> Honest manners are now reputed no more.*

Sebastian Brant's *The Ship of Fools* was a satire on the whole of society, especially the rulers of Church and state. Both these books ran to several editions within a few years and were translated into all the major European languages.

We are accustomed to applying the terms 'Renaissance' and 'Reformation' to this period of European history but the intellectual and the spiritual, the religious and the secular were so intertwined that it would be more accurate to think in terms of 'Re-valuation'. But the word, first coined in the nineteenth century, to describe the intellectual movement based on recovering the supposed virtues of classical civilisation was 'humanism' and its birthplace was Italy. This is scarcely surprising. Most of the refugees from the East who brought with them prized Latin and Greek manuscripts arrived first in Italy. They found in Rome and the city states of the north rulers who were (or liked to consider themselves) cultured and who patronised artists and scholars. What the members of these courts were now discovering was evidence of their *own* literary and intellectual heritage. Surrounded as they were by the ruins of an ancient civilisation, they now aspired to kick-start a cultural machine that had lain rusty and dusty for centuries. The movement they initiated was anthropocentric, hence the name later given to it.

> Humanitas ... opened a vista on the goals that could be reached on earth: individual self-development, action rather than pious passivity, a life in which reason and will can be

* S. Brant, *The Ship of Fools* (trs. Alexander Barclay, 1509), 'Prologue'

used both to improve worldly conditions and to observe the lessons that nature holds for the thoughtful.*

When we consider the revival of interest in nature our thoughts inevitably turn to Leonardo da Vinci (1452–1519). This Florentine polymath arrived about 1482 in the court of Ludovico Sforza, the ruler (and, from 1494, Duke) of Milan, a culture vulture determined to make his household the most glittering in Italy, if not in Europe. Leonardo's initial employment was as a musician but he was soon displaying his many other talents. He had an intense curiosity about everything he saw (or so it would seem from the notebooks he kept, which recorded details of landscapes, flowers, animals and, above all, the human form). He may truly be said to have made a science of art. The sixteenth-century art historian, Giorgio Vasari, recognised him as a pioneer 'modernist', capable of 'the subtlest counterfeiting of all the minutiae of nature exactly as they are'.† Leonardo made portraits of Ludovico and the luminaries of his court but he was soon going beyond the creation of fashionable likenesses. He studied skulls and bone structure. He made precise measurements of physiognomy to establish the proportions of the elements making up the head.

Leonardo must have begun his examination of human anatomy in his early days in Florence where, according to Vasari, the painter Antonio Polaiuolo (1453–1498) was the first to actually dissect human bodies. He later studied with Marcantonio della Torre (1481–1511), who lectured in anatomy at the universities of Pavia and Padua. The result was a collection of over 750 drawings, which were intended to be the illustrations for a treatise on anatomy. It was

* J. Barzun, op. cit., p. 44

† G. Vasari, *Le vite de' pi eccelenti, pittori, scultori e architettori* . . . (R. Bettarine and P. Barruchi eds.), Sansoni, Florence, 1968–1987, Volume 4, p. 8

never completed but one visitor to the artist's studio who saw the work in progress described it as including,

> . . . the demonstration in draft not only of the members, but also of the muscles, nerves, veins, joints, intestines and whatever can be reasoned about in the bodies both of men and women, in a way that had never yet been done by any other person.*

Leonardo was not an artist, pure and simple, whose miraculously precise observations were made so that he might better understand subjects he was called upon to paint. He was a philosopher, striving to comprehend the natural world and man's place in it. He pursued *scientia* for its own sake and, in this age of 'indispensable literacy', he planned to pass on his discoveries in written form. He read some of the Roman writers and early doctors of the Church. Like Augustine and other Christian logicians, he regarded the study of God's handiwork as a form of piety. In his dissection of the skull he pondered how visual images were recorded and understood. He conceived of the eye as connected to three ventricles or chambers, one for gathering data, one for processing it and one for storing it. The middle chamber, he surmised, was the location of the soul. But such observations and speculations could never absorb all his time and attention. He reflected on and filled his notebooks with ideas about optics, bird flight, mechanics, engineering and the movement of water. From first principles he calculated that it must be possible for man to travel in the free air like birds and in the airless world of fish. In the early years of what we call the Renaissance it was artists, rather than scientists, who expanded human knowledge and Leonardo was not the only genius to ask new questions about man and his environment.

* Cf. A.E. Popham, *The Drawings of Leonardo da Vinci*, The Reprint Society, London, 1952, p. 69

The main impetus for change, however, came from religious free-thinkers; earnest seekers after truth who felt impelled to challenge the Western Church and the dogmas that buttressed it. There had always been critics of the ecclesiastical status quo but the throbbing grid that carried power from the generator in Rome to monasteries, dioceses, universities and parishes throughout Europe had always proved more than equal to any challenge. Disciplinary procedures existed for dealing with unorthodox mystics and campaigners advocating reinterpretation of traditional teaching. As for those wilder spirits who rejected the embrace of Mother Church, ecclesiastic courts declared them to be heretics and the secular power could usually be relied upon to impose the death penalty upon them.

The vast majority of people, of course, raised no challenge to ancient custom. Some were genuinely devout. Others were too busy scraping together a living to bother their heads with matters of doctrine. Critics and sceptics there were who resented the behaviour of clerics whose motto appeared to be, 'Do as I say, not as I do.' The ethical demands of Christianity were so high that the failures of spiritual leaders attracted attention – then as now. As one lusty young Yorkshireman put it, 'Why should I confess an affair with a pretty woman to my knavish confessor who, given the chance, would use her similarly?'* Sometimes this protest against the power and privileges of the clergy took a more dramatic turn, as an incident in the 1520s suggests:

> Sir William Coffin ... passing by a churchyard ... saw a multitude of people standing idle; he enquired into the cause whereof; who replied, 'They had brought a corpse thither to be buried but the priest refused to do his office, unless they first

* A. G. Dickens, *Lollards and Protestants in the Diocese of York 1509–1558*, OUP, Oxford, 1959, p. 245

> delivered him the poor man's cow, the only quick goods he left, for a mortuary.' Sir William sent for the priest and required him to do his office to the dead; who peremptorily refused it, unless he had his mortuary first. Whereupon he caused the priest to be put into the poor man's grave, and earth to be thrown in upon him; and he still persisting in his refusal, there was still more earth thrown in, until the obstinate priest was either altogether, or well nigh suffocated.*

The monolithic Church exercised temporal power through its vast land holdings, its courts and its disciplinary procedures but its power lay not only in the temporal realm. The sacraments and penitential rites existed to ease the soul's passage from this world to the next, even if few could afford the liberality of Philip II of Spain (died 1598), who ordered that a mass was to be celebrated in the monastery of the royal palace the El Escorial 'every day until Christ's second coming'.

Making intercession for the departed was not confined to the 'Church Militant' (i.e. the company of believers in this world): the Pope and his priestly subordinates could also call upon the prayers of the Church Triumphant. Christ and the saints in heaven had, by their prayers and holy actions, accumulated a 'treasure of merit' upon which Christians could draw when pleading their case before the divine Judge. In this way the Catholic Church controlled all members of Western Christendom, body and soul. It is not difficult to understand that such power could provoke resentment but also fear: few were bold enough to challenge the teaching of the Church when such rebellion might have terrifying eternal consequences.

By 1500 the need for reform was obvious to all thinking people except die-hard reactionaries. It was not just a matter of isolated

* J. Price, *Danmonii Orientales Illustres: or, The Worthies of Devon*, Rees and Curtis, London, 1810, p. 228

local grievances; moral laxity and corruption infected the ecclesiastical establishment from the top down. Under the papacies of Alexander VI (Rodrigo Borgia), Julius II (Giuliano della Rovere) and Leo X (Giovanni de Medici) between 1492 and 1521, the Vatican plumbed hitherto unprecedented depths of venality. The Popes were politically preoccupied with preserving and extending their territorial possessions in Italy and using their prestige to win the support of European monarchs. This ambition was one contributory cause of the intermittent Italian Wars (1494–1526), which involved France, the Holy Roman Empire, the Italian city states and (to a lesser extent) England. Military and diplomatic priorities forced religious reform well down the agenda.

Unfortunately for the holy fathers, it was at this precise time that intellectual questions were being posed that urgently required addressing. The most pressing came from new ways of reading and interpreting the Bible. As Professor MacCulloch has observed, 'Even if the events of the 1520s had never happened and there had been no evangelical challenge to the Church, the coming of printing would have changed the shape of religion.'* For more than a thousand years before the 1450s – when the invention of printing with movable type made possible the publication and dissemination of vernacular Bibles – the text had been the exclusive preserve of scholars able to read, discuss and teach Jerome's Latin version, the Vulgate. It consisted of seventy-six Old Testament books, forty-six New Testament books and three Apocrypha.

Since the whole Vulgate embraced various types of literature – history, poetry, philosophy, prophecy, law, letters and the Gospel accounts of the life and teachings of Jesus – it was always obvious that different exegetical criteria had to be brought to bear upon any text before it could be used for personal devotion or public teaching.

* D. MacCulloch, op. cit., p. 73

But students did not attain to the dizzy heights of theology until they had mastered the seven liberal arts of grammar, logic, rhetoric, music, astronomy, geometry and arithmetic. Only then were they considered mentally equipped to wrestle with the sacred text. Except that they did not actually come to grips with it personally. The Bible is so complex and it was important that students should interpret it *correctly* – i.e. according to the approved traditional meaning; so they handled it with the kid gloves of the accumulated wisdom of the great doctors of the Church.

This wisdom took the form of 'glosses'. Originally these were marginal notes providing explanations of difficult passages but they developed into running commentaries. In the mid-twelfth century, Peter Lombard, the Bishop of Paris, had published the four *Books of Sentences*, a monumental mega-commentary that combined biblical texts with quotations from the Church Fathers, covering all major aspects of Christian doctrine. This became the fount of orthodoxy and the essential textbook for all theology students. The *Sentences* was the bedrock of intellectual unity across Western Christendom, ensuring standardisation of belief and biblical interpretation.

Scholasticism fixed the way the various books of the Bible were to be read and expounded. The foundational truth was that they were all divinely inspired and, therefore, all in agreement about God's plan in creation and redemption. In some places (e.g. the Gospels) that plan was more obvious than in others (e.g. the Song of Songs). The reader needed the guidance of the Holy Spirit and the intellectual training to apply different methods of exegesis to the various strands within the fabric of the holy text. The perfected system of study infallibly provided this – or so the masters in Europe's theology schools claimed. God had appointed the Pope as both the guarantor of truth and the guardian of correct methods of biblical interpretation.

There were various ways in which any passage of Scripture might be approached. The literal or historical approach meant taking the words at face value. The typological approach was a means of discerning Christian truth within the pages of the Jewish writings (the Old Testament). Thus, for example, Noah's Ark became a picture of the Church, God's chosen vehicle for saving mankind from the devastating flood of the Last Judgement. Allegorical treatment superimposed spiritual understanding onto a straightforward reading of the text, as in discerning Eucharistic significance in Jesus' first miracle of turning water into wine. Tropological exegesis focused on the moral lessons to be drawn from biblical incidents. Anagogical reading pointed away from this world to events in the heavens, particularly to the eternal destiny of the Christian soul.

This way of dealing with real or apparent difficulties in the biblical text became self-defeating. By the sixteenth century the flaws in the convoluted approach were becoming obvious to impatient teachers. Many agreed with the outstanding Swiss preacher, Geiler of Kaisersberg, who scornfully dismissed figurative interpretation as a method that had made the Bible a 'nose of wax' to be turned in whatever direction took the speaker's fancy. Humanist theologians wanted to treat it in the same way as any other ancient text. This did not mean that they denied its divine inspiration – far from it. They were concerned to let it speak for itself, believing that it would speak more convincingly if it was freed from theological accretions.

There were compelling reasons for a rethink about the Vulgate. Innumerable copyists down the centuries had introduced errors. Also the discovery of previously unknown early texts made it possible to do a rewrite that would be closer to the original. Stylistically, Jerome's work could be improved upon thanks to the newfound availability of some of the elegant Latin writings from the golden age of classical literature. More than all this, European scholars were becoming more adept in the languages of the Hebrew and Greek originals.

All in all, scholars now had the opportunity to produce a much improved, standard Christian text and one which, thanks to the printing press, would be free from the hazards of hand-copying.

But did the ecclesiastical hierarchy want it? It was a fundamental belief that the Church had, under the guidance of the Holy Spirit, created the Bible. Could it now accept the implied criticism that it had not made a very good job of it? There was certainly unease among some defenders of the status quo. The most notorious arguments concerned the *comma Johanneum.* In 1516, Desiderius Erasmus (*c.* 1466–1536), the internationally celebrated classical scholar, published his *Novum Instrumentum Omne*, a version of the Greek New Testament based on the best documents to which he had access. It was very quickly noted that he omitted a phrase (called a comma) occurring (in Latin) in the Vulgate at I John 5,7-8. The whole passage reads:

> There are three that bear record *in heaven, the Father, the Word and the Holy Spirit and these three are one. And there are three that bear witness in earth*, the spirit and the water and the blood . . .

The italicised words are those omitted by Erasmus. When challenged, he responded that he could not find the comma authenticated in any early document. Now, these words clearly and emphatically proclaim the doctrine of the Holy Trinity and they are the only words in the Bible that do so. It was in the interests of the Church authorities to defend their authenticity. They did so vigorously and soon the fur was flying, with accusations of fraud and intellectual incompetence being thrown around. Erasmus eventually yielded to pressure. In a later edition, he reinstated the missing words, but only under protest. Controversy has continued down to the present day and among those who became involved was Isaac Newton (see p. 243).

In the early stages of what we call the Reformation, the Vatican did not grasp the significance of what was, in fact, a *scientific* challenge to its authority. If linguists (and Erasmus was not alone) were asserting that the Bible, like any other early text, was subject to scrutiny by specialist scholars, then the claim of the papacy to be the only arbiter in all matters of Holy Writ was being challenged. Authority was being shifted from Christ's vicar to the linguists. And worse was to come.

> I vehemently dissent from those who would not have private persons read the Holy Scriptures nor have them translated into the vulgar tongues, as though either Christ taught such difficult doctrines that they can only be understood by a few theologians or the safety of the Christian religion lay in ignorance of it. I should like all women to read the Gospel and the Epistles of Paul. Would that they were translated into all languages so that not only Scotch and Irish, but Turks and Saracens might be able to read and know them.*

Erasmus boldly proclaimed this in the preface to the first edition of his *Novum Instrumentum* and he reiterated the view more forcefully in later editions. So, every man – and every woman(!) – was to become his/her own interpreter. What was particularly disturbing to the Catholic hierarchy was that this distinguished and highly respected scholar was aligning himself with heretical tinkers, weavers and humble artisans who gathered in secret to read unauthorised versions of the sacred text. As early as 1408 the Archbishop of Canterbury had ordered:

* P. Smith, *Erasmus – A study of his Life, Ideals and Place in History*, Frederick Ungar Publishing Co., New York, 1962, p. 184

> That no-one henceforth on his own authority translate any text of Holy Scripture into English or any other language by way of a book, pamphlet or tract and that no book, pamphlet or tract of this kind . . . be read in part or in whole, publicly or privately, under pain of the great excommunication, until the translation shall have been approved by the diocesan of the place.'*

What had provoked this blanket proscription was the English Bible written by John Wycliffe and his aides and copied by devoted followers, or 'Lollards'. Interestingly, the same archbishop had earlier sanctioned an English version of the four Gospels for the wife of Richard II – a version that had almost certainly come from a Wycliffe-derived source. It would seem that the authorities were caught in two minds. They had a long history of dealing drastically, often violently, with heretical groups. They were ready to hand men and women over for burning merely for owning an unauthorised Bible, but they were less enthusiastic about exposing the errors they insisted were contained in the illicit volumes. There was a reluctance to look critically at the text which would have meant taking seriously the issues Wycliffe and others raised. The papacy and the untrammelled word of God could not possibly be in conflict. Defenders of the papacy simply shouted down the opposition.

There was no official papal line on translation of Scripture into languages 'understanded of the people' – at least, not until it was too late. Attitudes within the Catholic establishment had been mixed. From time to time psalms, Gospels and reflections on biblical themes and passages had been produced as popular aids to devotion. Many welcomed printing and the impetus it gave to literacy as a means of informing and deepening conventional faith. At the same time, master printers were quick to sense profit. It was only natural

* A. W. Pollard, *Records of the English Bible*, OUP, Oxford, 1911, p. 79

that many people would want to read for themselves the basic handbook of the Christian faith. There was an enormous demand for the word of God among intelligent, non-Latin speaking people. Whole and part-Bibles were among the first books to come off the new presses. Vernacular Bibles were being printed in Germany by 1466, in Italy by 1471, in France by 1473, in the Low Countries by 1477 and in Spain and Bohemia by 1478. Their impact was volcanic.

But the Bible was not the only text being avidly studied by the intellectual avant garde. Between 1490 and 1506 a complete edition of the works of St Augustine was published in Basel. Readers could now study the great doctor of the Church free of later interpretations. They discovered his emphasis on grace and inward faith as opposed to reliance upon rituals and priestly intercession. A similar emphasis was stressed by the French humanist, Jacques Lefèvre d'Etaples, in his commentaries on the epistles of St Paul and, in 1504, Erasmus, himself, produced a manual of Christian discipleship, the *Enchiridion Militis Christiani* (*The Christian Soldier's Dagger* – or perhaps a more accurate modern translation might be 'trench tool'). This hugely popular work was a handbook of lay piety that dismissed the supposed superior spirituality of the monastic vocation and urged the importance of direct communion with Christ based on meditation on Scripture.

If individualism was not itself invented in the age of the Renaissance/Reformation, in Erasmus' call for universal reading of the Bible it may be said to have come of age. If everyone was free to read the sacred oracles of God and determine his/her own response, becoming entirely in charge of his/her eternal destiny, without the need for the prayers of the saints or the rites of the Church, then the foundations of papal power would crumble.

The vast majority of Erasmus' contemporaries, of course, stuck to the old familiar patterns. They attended mass, made their confessions, went on pilgrimages, venerated relics and believed as their priests told them to believe, but vernacular Scripture was a crowbar that had already been thrust into the crack of Catholic solidarity.

The man who most effectively applied pressure to that tool was Martin Luther (1483–1546) and his challenge went to the very heart of the soteriological problem; the issue which stirred in the hearts of many people that resentment and fear of which we have spoken. Did the Pope and his cohorts really have power over the souls in purgatory?

What began the conflict of the Reformation was Leo X's issuing of a new, Europe-wide indulgence to raise funds for the rebuilding of St Peter's Basilica in Rome. There was nothing new about the indulgence system, as we noted at the beginning of this chapter. It was a recognised source of income, not only for the Church, but also for temporal rulers whose projects had its blessing. For example, the Elector Frederick of Saxony issued certificates to pilgrims who came to view his large collection of holy relics. This provided an answer to the question, 'What must I do to be saved?' But only *if* the treasury of merit really existed, *if* the Pope had access to this treasury and *if* works of pious charity (such as contributing to Leo's building fund) could earn eternal merit for the performer. But these were big 'ifs' and Luther was not the first to raise them. And there was something particularly audacious about Leo's indulgence. Those hawking it – the most notorious of whom was the Dominican, Johann Tetzel – were claiming that this papal magic spell had the power to release from purgatory the souls of the dear departed. In a word, the indulgence trade had become a racket.

Martin Luther was a monk in the Augustinian cloister at Wittenberg in Saxony and a lecturer on the Bible at the university. Like countless devout men and women before him, he had taken the cowl in order to discover the answer to the question, 'What must I do to be saved?' Striving to achieve the absolute holiness demanded by a perfect and just God plunged him ever deeper into despair until his close study of Paul's letter to the Romans made him reflect on the statement, 'God's righteousness [justice] is revealed from faith to faith, as it is written "the righteous [justified] men will live by faith".'*

The implication of the conviction that God requires only a response of faith – trust, commitment – was that all the penitential paraphernalia by which good Catholics attempted to buy their way into heaven was useless. Now, 'salvation by faith alone' (*sola fide*) was not a new idea. It had been central to Augustine's theology. Other Church doctors had taught it. It could be found in Peter Lombard's *Sentences*. But the problem lay precisely in those words, 'could be found'. Justification by faith was part of the vast corpus of orthodox Christian teaching but it was one tree in a forest, obscured by the luxuriant scholastic growth all around it. What should have been a towering pine immediately visible to the searching eye was hidden from view by a Church that had been too eager to present an all-embracing, all-sufficient corpus of doctrine.

The more Luther studied the Bible, the more he discovered aspects of official teaching and practice that seemed to be at variance with the text. It was his challenge of the indulgence traffic in 1517 that thrust him into the spotlight. He did not intend to challenge papal authority or show it up as in opposition to the word of God (he was careful to suggest that Pope Leo X's initiative had been perverted by corrupt agents). He had other matters on his mind in 1518 and 1519, principally preparing and delivering a course of lectures against Aristotle, whom he loathed as a 'blind pagan' and 'a mere Sophist and quibbler'. Instead he called for a scholarly debate on the indulgence question and proposed ninety-five points for consideration.

No one took up the challenge, but printers, seeing profit in this sensational document, were soon churning out copies that

* Romans 1:17

were enthusiastically read and circulated. An academic exercise became a *cause célèbre* because it touched an exposed nerve of anti-clericalism; resentment of the power of clergy and the interference of the papacy in affairs north of the Alps. Even so, the whole affair might have been no more than a storm in a teacup had the Vatican not reacted so violently and if Luther's own patron, the Elector of Saxony, Frederick III ('the Wise'), had not extended his protection to the monk instead of delivering him up to papal justice.

In 1520, Luther was ordered to appear before representatives of the Pope and emperor to give an answer for his 'heretical' opinions at the imperial diet held in Worms. When ordered to withdraw his writings, he replied, 'Unless I am proved wrong by Scripture and plain reason, I cannot and I will not recant.'

What did he mean by 'plain reason'? Well, certainly the literal understanding of the text, stripped of all allegorical and typological accretions. Probably he also had in mind the glosses adorning every page that told the reader how to interpret the text. But he recognised that the content of the Bible had been decided by the early Fathers who had categorised the various writings into the canon and apocrypha and had named a third group 'antilegomena', disputed writings falling between the other two groups. Although not rejecting any of the books of Jerome, Luther relegated to the end of his Bible Hebrews, James, Jude and Revelation. James met with his particular disapproval and was dismissed as the 'epistle of straw' because it appeared to be at odds with St Paul's insistence on justification by faith, in that it stressed the importance of performing good works.

Luther was certainly guilty of inconsistencies in his approach to interpretation. In disputing with Erasmus he stated emphatically, 'When we show ourselves disposed to trifle even a little and cease to hold the sacred scriptures in sufficient reverence, we are soon involved in impieties and overwhelmed with blasphemies.'* Yet, in his own translation of Romans 3:28 – 'A man is justified by faith without works of the law' – he did not hesitate to add 'alone' after the word 'faith' in order to emphasise the central doctrine of *sola fide*.

Since German Bibles had by this time been freely available for half a century it is pertinent to ask why other readers, before Luther, had not made the breakthrough into a personalised, evangelical (i.e. 'gospel') faith. The rapid and widespread response to the challenge from Wittenberg would, surely, suggest that *sola fide* was an idea whose time had come. The answer might simply be that evangelicalism needed a hero, a champion, a leader – someone prepared to light the blue touch-paper. Yet, I suspect there was more to the timing than that.

One problem was that the existing vernacular versions did not exactly make riveting reading. They were plodding translations of the Vulgate, heavy with official glosses. They were vehicles for reinforcing orthodox doctrine. If the mood of the times was being changed by the increasing number of literate people, it was thanks to other kinds of books, including devotional manuals and pamphlets. They may not have been questioning orthodox teaching but they were doing something yet more profound: they were encouraging people to reflect on their faith *individually* and from the *printed word*. And there were sceptical, disrespectful, even scurrilous books, such as the runaway bestseller, *The Ship of Fools* and Erasmus' equally popular satire *In Praise of Folly* (1511). The author claimed to be surprised and embarrassed by the popularity and re-translation of the gentle ridicule of the Establishment. These were only the two front-runners in a race in which several sceptics participated. Criticism and resentment of ecclesiastical power (or rather its misuse) was not new. It went back earlier even than the time of

* Martin Luther, *The Bondage of the Will*, trs. J. Packer, O. Johnston, James Clarke, Cambridge, 1957, p. 85

Chaucer. But, by 1517, there was a widespread mood in society that went beyond amused disrespect. What Luther did was lend popular discontent a theological legitimacy.

Having escaped his enemies at Worms, Luther exchanged furious fire with his enemies at Rome, which he now identified with Babylon, the seat of Antichrist, portrayed in the Book of Revelation. But his most important publication was his translation of the Bible into High German, a work which, as literature as well as scholarship, was a triumph. But did Luther set before his growing band of followers an unadorned text they could interpret for themselves? He did not. Both the New Testament (1521) and the Old Testament (1534) carried introductions to the various books and marginal glosses indicating how the text should be interpreted. Luther's Bible was 'modern' in shaking itself free of the Vulgate and in being based on the best early Hebrew and Greek texts but in its method of presentation it was medieval. We have already seen how he 'helped' St Paul explain his theology of saving faith by adding the word 'only' to the text of Romans. As time passed and Luther came into conflict with other scholars – Protestant as well as Catholic – he displayed a stubbornness and even an inconsistency in his interpretation of Scripture in the light of 'pure reason'.

The leaders of the main evangelical groups that emerged over the next twenty years were agreed on the fundamentals but developed differences of emphasis over which they contended with each other as earnestly as they did with the Roman Antichrist. Luther and Erasmus clashed in their attempts to square the circle of divine predestination and human free will. Luther repudiated the iconoclasm of radicals who defaced church statues and paintings in strict obedience to the command, 'You shall not make any graven image'. More crucially, he was at loggerheads with Ulrich Zwingli, leader of the reform movement in Zurich over the understanding of the Last Supper. Zwingli believed the words of institution, 'This is my body', should be interpreted figuratively while Luther insisted that they should be taken literally. This was more than a theological quibble. It fatally undermined an attempt at the Colloquy of Marburg (1529) to unite evangelicals and ensured that the Protestant camp would remain divided.

Such discord was music to the ears of Catholic critics, who saw this as proof that leaving Mother Church led to doctrinal chaos. This, in turn, obliged evangelical education to do precisely what Catholic teachers had done before them: set out in great detail how the Bible was to be interpreted. The king of the Protestant systematisers was Jean Calvin (1509–64). This French humanist who, in 1533, fled from persecution in Paris, settled in Geneva and eventually turned that city into the Protestant Rome. He knew that the evangelical world needed a theological framework as comprehensive and rigid as that of medieval Catholicism, and spent twenty-four years between 1535 and 1559 developing an extensive Bible-based *scientia* – *Christianae Religionis Institutio,* commonly known as the *Institutes.* It began, 'Nearly all the wisdom we possess . . . consists of two parts; the knowledge of God and of ourselves.'

Fundamental to Calvin's thinking was the sovereignty of God. No one can speak for the Creator – certainly not the Pope. All we can know of him is to be found in what he has revealed in his written word. Beyond that we may not speculate. Wisdom entails a realisation of the limits of reason; an acknowledgement of our profound ignorance. This was the lesson of the closing chapters of the book of Job. The much-suffering servant of the Lord, seeking the reason for his unmerited distress, was finally answered by God:

> Who are you to question my wisdom with your ignorant,
>   empty words? . . .
> Were you there when I made the world? If you know so
>   much, tell me about it.
> Who decided how large it would be . . . Do you know all the
>   answers? . . .

> Who laid the cornerstone of the world?
> In the dawn of that day the stars sang together and the heavenly being shouted for joy.*

Thus the divine answer rolls on through more than a hundred verses of magisterial poetry. And there Calvin was content to leave the matter.

Early reformist scholars were not opposed to enquiry into the physical workings of the universe but they shared with their Catholic counterparts opposition to any theories that appeared to conflict with the words of Scripture, plainly understood. Both Luther and Calvin were scornful of heliocentrism but they did not (as some later commentators suggested) condemn Copernicus as a heretic. Indeed, the Lutheran mathematician, Rheticus (Georg Joachim de Porris), a disciple of Copernicus and a pioneer of trigonometry, came to his master's aid with a defence in which he cited several Bible references in which the text seemed to be at variance with later knowledge about the cosmos. He insisted that the biblical writers described the *appearance* of phenomena, rather than their actual character. However, he seems to have chickened out of making his views widely available; no extant published work on the subject is known prior to the mid-seventeenth century.

One practice Luther and Calvin were both clear in condemning was judicial astrology, the prophecy of specific events foretold in the stars.

> There has been for a long time a foolish curiosity which consists of judging by the stars all that should happen to men, and of enquiring of them what course to take . . . Rejected in the past as pernicious to the human race this phenomenon is in full revival today, with the result that many people who

* Chapter 38, 2–7

> believe themselves to be of sound mind and who indeed have the reputation of being so are almost bewitched.*

So Calvin wrote in a little treatise on the subject. His opposition was based not on an analysis of metaphysical argument, but purely on theological objection. Since God, by the exercise of his sovereign power, determines the fate of every one of his creatures, and since he seldom reveals his purposes in advance, predictions based on horoscope readings were not only inaccurate, but impious. It followed naturally to Calvinists that astrology was a deception practised by the devil.

Yet in the hot and turbulent climate of the Reformation, when old beliefs, customs and attitudes were being challenged, astrology still held sway over many minds, including some close to the centre of the Protestant revolt. Luther found himself at odds with his friend and devoted assistant, Philip Melanchthon, who, like most contemporary thinkers, wanted to find a place for astrology within the realm of science. Among the off-the-cuff remarks recorded by Luther's students and later published as *Table Talk*, there are references to the reformer's rejection of Melanchthon's arguments:

> My Philip has devoted much attention to this business, but he has never been able to persuade me . . . He himself confesses, 'Though there is a science in it, none has mastered it' . . . I replied, 'Astrologers are silly creatures to dream that their crosses and mishaps proceed not from God, but from the stars.†

* J. Calvin, '*Avertissement contre l'astrologie qu'on appelle judiciaire*', cf. B. Cottreet, *Calvin – A Biography*, John Knox Press, Kentucky, 2003, p. 6

† *Luther's Works*, ed. H. T. Lehmann and J. Pelikan, Concordia Publishing House, Missouri, 1955 – volume 54, p. 173

We can look at the evangelical churches that had emerged by the turn of the seventeenth century and see them as a lurid and bewildering kaleidoscope of clashing colours and shapes (see p. 168). Alternatively, we can discern patterns and themes that bind the disparate parts into a unity. Central to all the teachings is the Bible, plainly understood. But that understanding is the responsibility of the individual, for, although the different churches had their own disciplinary structures and were ready to excommunicate members who wandered from the approved doctrines of the group, they all proclaimed the same answer to the old question, 'What must I do to be saved?' That answer was, 'Believe in the Lord Jesus Christ and you will be saved.' The individual's eternal destiny was in his/her own hands. One did not become a Christian by belonging to the Church; one belonged to the Church by becoming a Christian.

The intellectual backgrounds of the reformers were varied. Their thought habits had been formed in different schools. For example, Calvin, the humanist, had a wide and deep knowledge of the ancient Greek and Latin authors and quoted them freely in the *Institutes*. Luther, by contrast, found the old philosophers worse than useless. He struggled to understand their quibbles. When obliged to lecture on philosophy early in his career, he wrote to a friend, 'I would gladly have exchanged it for theology; I mean that theology which searches out the nut from the shell, the grain from the husk.'*

Yet both men arrived at what was more than an intellectually satisfying corpus of ideas; it was a faith to live by – and, if necessary, to die for. Their lives were illumined by reason, revelation and hope. Reason was the gift God had supplied to enable his creatures to understand their world. Revelation was God's gift of enlightenment, imparted primarily through faith in the scriptures, which enable

* Cf. G. Rupp, *The Righteousness of God – Luther Studies*, Hodder & Stoughton, London, 1953, p. 93

men to know him, insofar as that was possible in this world. Hope was the gift by which they accepted what they did not and could not know, being content with the limits of their knowledge. They faced the question that had been presented to Job: 'Can you by searching find out God?' And they averted their gaze.

Rome, of course, had not been inactive while all this was going on. The first response of Church authorities to the growing number of vernacular Bibles and the critiques based on them was censorship. Lists of banned books were issued in various countries, beginning in the Netherlands in 1529. It was another thirty years before a complete *Index Librorum Prohibitorum* was published in Rome, by which time it had anathematised the works of 550 authors, including, of course, vernacular Bibles. Proscription was not restricted to religious texts; the fact that an author, however learned, was a Protestant was sufficient to get his work placed on the Index. Catholic scholars were – theoretically at least – denied access to the writings of their peers – books such as Sebastian Münster's great world atlas, *Cosmographia*, and the treatises of the pioneer botanist, Otto Brunfels.

The Vatican wrestled with a mounting task of vetting new works. Scores of Bible translations were now in existence all over Europe, many carrying tendentious marginalia or woodcut illustrations. In England alone, six royally approved versions appeared between 1537 and 1611. In 1571, the work of censorship was put on a well-organised, permanent footing with the establishment of the Sacred Congregation of the Index. Its members met twice a year to update the list and were kept busy the rest of the time hunting down and reading all suspect titles. They had power to burn books and close down printworks.

At the Council of Trent (1545–1563), where the Catholic Church set about systematically putting its house in order, nothing caused the assembled cardinals and bishops more problems than the Bible. There was no chance now that mere proscription would rid the world of Protestant versions; the Vatican would have to engage in the business of re-education. Still, conservatives stuck to their guns over the Vulgate, insisting that it was the only version preserved by the Holy Spirit from error and demanding that any modern translations must be based on Jerome's original. Furthermore, it was decreed that only devotional books of the Bible, such as the Psalms and the Gospels might be reworked. The epistles of Paul and the Revelation of St John contained material that was much too 'hot'. The Catholic hierarchy had to take leaves from the enemy's book. They insisted that any new translations should carry anti-heretical marginalia and they encouraged the setting up of lectureships on Scripture. This 'if you can't beat them, join them' attitude was an acknowledgement that, in many areas, the Protestant churches were winning the battle of the pulpit. Luther's insistence on '*Sola Scriptura*' – solely relying on Holy Scriptures – had replaced an image-based Christianity of holy artefacts and symbolic acts with a word-based religion that demanded the instruction of the faithful in doctrine and encouraged them to *think* about what they believed. One man who was acutely aware of the problem that Catholic clergy were not engaging with the plain word of God was William Allen, an exile from Elizabeth's England. He established at Douai a college (later moved to Rheims) to train priests to send back as missionaries to his homeland.

> Catholics educated in the academies and schools have hardly any knowledge of the scriptures except in Latin. When they are preaching to the unlearned and are obliged on the spur of the moment to translate some passage into the vernacular, they often do it inaccurately and with unpleasant hesitation because either there is no vernacular version of the words or it does not occur to them at the moment. Our adversaries, however, have at their fingertips from some heretical version all those passages of scripture which seem to make for them

> and, by a certain deceptive adaptation and alteration of the sacred words, produce the effect of appearing to say nothing but what comes from the Bible. This evil might be remedied if we too had some Catholic version of the Bible, for all the English versions are most corrupt.*

So Allen complained in 1578. The eventual result was the Rheims-Douai Version that appeared in 1582 (New Testament) and 1609–10 (Old Testament). This work slavishly followed the Vulgate, sometimes to the point of offering obscure Latinisms, but it did have some compensating virtues and the committee working on the King James Bible did not hesitate to take note of the new version. This royally sanctioned text would remain the standard English Bible for four centuries.

* *Letters and Memorials of Cardinal Allen* (1882), trs. H. Pope, 1952, p. 250

CHAPTER FOUR

Heavens above?

In the kaleidoscope of sixteenth-century truth-seekers there were some scholars who had little interest in metaphysics – men we might more easily identify as what we now think of as 'scientists'.

The most obvious of Cornelius Agrippa's contemporaries to fall into this category was Nicolaus Copernicus. He was born in 1473 into a wealthy Polish family with important national and international connections. As a young man he took minor Catholic orders and never contemplated challenging Church doctrine. Family influence and his own talent raised him to the prominent position of secretary to the Prince Bishop of Warmia. As such he was involved in the political, religious and diplomatic life of his powerful employer. What made Copernicus different from other high-ranking officials was his powerful intellect. He was a polymath who, having studied at various leading universities, emerged with a sound understanding of the humanities as well as qualifications in law and medicine.

His mind was engaged above all else by mathematics and astronomy. He was fascinated by the movements of heavenly bodies and particularly in those movements that did not appear to 'obey the rules'. According to the Ptolemaic system which had, for centuries, been accepted as providing the map of the heavens, the Earth was the centre of a universe of concentric circles in which the sun, moon and known planets of the solar system rotated within an outer circle of stars and astral entities whose positions were fixed.

In the generations before the invention of the telescope, knowledge of the universe could only be gained by measuring the movements of all observable bodies in relation to each other. The problem that intrigued stargazers was that these movements were not constant and did not always agree with the courses projected for them by Ptolemy. One explanation that made sense of such seemingly erratic behaviour was that terrestrial measurements were not made from a fixed point; that the Earth, too, was a planet in motion around its own star, the sun.

Copernicus did not invent the theory that later came to bear his name. The possibility of a heliocentric solar system had occurred to Greek and Arabic scholars in classical times. The problem was that it appeared to fly in the face of common sense. If mankind inhabited a lump of rock whirling through space, why was there no wind constantly ripping across its surface and how could objects not anchored to the ground avoid being torn off by their force? In addition, objections were made based on certain biblical references. Since the Jewish and Christian scriptures had been written from within a terracentric world view, they apparently endorsed the conviction that the Earth was the focus of God's attention, that the sun, moon and stars had been created for man's benefit, and that the Creator lived in 'heaven', a realm beyond Ptolemy's spheres.

Copernicus may not have originated the heliocentric solution but he did subject it to closer investigation than anyone else. He began his observations, measurements and calculations around 1510 and, over the ensuing years, carried out thousands of viewings of various heavenly bodies. He discussed his findings confidentially with a small circle of sympathetic scholars and he tentatively published his conclusions, though it would not be until 1543, at the age of seventy, that he fully explained his theory in *De Revolutionibus Orbium Coelestium* (*On the Revolution of Heavenly Spheres*). This suggested not only that the Earth and its sister planets revolved around the sun, but that the entire cosmography of the heavens was vastly different from that which had been previously assumed. The firmament (Ptolemy's outer ring) was inconceivably distant.

Copernicus' system was eagerly debated by astronomers, mathematicians and theologians – largely in a tone of civilised discourse. What the Polish scholar was presenting for debate was an elegant solution that neatly explained the secrets of heavenly movements. What he was saying was, in effect, 'If my calculations are correct, this is what the universe looks like.' Copernicus had several disciples and more than a few admirers – including some in the Catholic hierarchy. Inevitably, he also had critics. For some scholars his solution was too neat. Men steeped in Aristotelian physics, with its insistence on experimental proof, accused him of putting the cart before the horse. 'We observe the universe to look like this,' they insisted. 'Therefore, your calculations are not correct.' Thinkers not seduced by the 'mystique' of numbers regarded mathematics as merely an intellectual exercise. At best it was a useful tool but it could not replace physical evidence.

Copernicus died days before his major work hit the bookstalls, enabling friend and foe alike to respond in detail to his thesis. What is interesting is that there was no move by reactionary churchmen to have this 'heretic's' bones dug up and consigned to the flames or to make bonfires of his writings. Such disputation as occurred proceeded from the lecture hall and the print shop. If Copernicus had delayed full publication until the last possible moment for fear of giving offence, it seems that he need not have worried.

Several possible reasons offer themselves for the relatively tame reception accorded to this seminal work. The first is that the author did not adopt a belligerent stance to received wisdom. Copernicus was no loud-mouthed revolutionary with a massive ego, determined to draw attention to himself by throwing down the gauntlet to the scholarly establishment. He was a son of the Church with impeccable ecclesiastical credentials. For the Catholic hierarchy he was 'one of us'. Moreover, he never drew any philosophical or religious conclusions from his findings. He simply brought fresh thinking to what had long been a *scientific* problem. Nor did Copernicus offer any observations on astrology – a subject in which, apparently, he had no particular interest.

Insofar as his theory presented a challenge to the Bible it is important to recognise that not all scholars were wedded to strict literalism. It was widely understood that the writers used poetry and other literary forms, that the biblical narrative was about God's work in salvation and not a textbook describing the workings of the created order, and that some books reflected the understanding of the cosmos at the time of composition. Augustine, for example, had pointed out that Isaiah's reference to God's spreading out the firmament like a curtain (Isaiah 40:22) could not be regarded as contradicting what Augustine's educated contemporaries believed – that the heavens were spherical.

However, a more compelling explanation for the relatively muted reaction to heliocentric theory was the prevailing mood of the times. It was in 1529 that the word 'Protestant' first made its appearance. It had a very restricted use, applying only to a group of German princes and urban leaders who embraced the beliefs of the reformers Martin Luther and Huldrych Zwingli and resisted pressure from the Catholic Holy Roman Emperor. Only gradually did 'Protestant' and 'Catholic' become badges worn by individuals and – more significantly – by nation states. At the time of Copernicus' death the cracks of theological stress had not yet widened into deep fissures snaking their way through Europe, dividing the continent into bitterly opposed politico-religious camps that would be at war with each other, on and off, for more than a century.

Religious belief changed the whole nature of warfare. From the 1560s onwards armies clashed not simply at the behest of territorially ambitious rival princes nor even to assert national supremacy. War took on the nature of holy crusade – crusade not between defenders of rival religions, but between partisans of rival versions of Christianity.

The individual foot soldier or mounted knight believed that he was blood-letting for sacred and eternal truth. And the men who defined that truth were preachers, priests and scholars.

What individuals believed and taught was now more important than ever before. Theological and philosophical minutiae became principles to fight and die for. Scholarship could not be divorced from politics any more than politics could be divorced from religion. As ideological conflict intensified, academic courtesies declined. It follows that we cannot understand the changing world of scholarship unless we look beyond the lecture halls and publishing houses to see what was happening in the wider world. A catalogue of the major events of just one decade dramatically makes the point:

1580 The Seventh War of Religion broke out in France; Francis Drake's return from his circumnavigation of the globe inaugurated the era of England's colonial rivalry with Spain and Portugal.

1582 Pope Gregory XIII ordered the universal observance of the Gregorian calendar.

1583 Plots to invade England and assassinate Elizabeth I were thwarted.

1584 William of Orange, Protestant leader in the United Provinces, was assassinated.

Nicholas Sanders published *De Origine et Progressus Schismatic Anglicani*, the Catholic martyrology that became the foundation text for Counter-Reformation in England.

1585 Elizabeth I reluctantly became protector of the Protestant United Provinces and sent troops to the Netherlands.

1586 Further plots against Elizabeth I were unmasked.

1587 Mary, Queen of Scots was executed (having named Philip II of Spain as her heir).

1588 Philip II's naval invasion of England failed; the Guise brothers, leaders of French Catholics, were assassinated.

1589 Henry III of France was assassinated.

Including the establishment of the Gregorian calendar in this dismal catalogue may seem strange but it (or rather the reaction to it) was one of the most disruptive elements in Europe for over a century. Protestant states staunchly resisted abandoning the old Julian calendar. The eleven-day discrepancy between the two reckonings brought confusion to diplomatic and business activities and for record-keeping in general (not to mention making life difficult for later historians). By 1600, confessional differences were felt at all levels of society and added bitterness to the relationships between states. This mutual hostility could not fail to affect scholarly discourse. Just how belief and *scientia* could lead devotees along dangerous, though different, paths can be illustrated by the lives of three close contemporaries, John Dee, Tycho Brahe and Giordano Bruno.

Dee was born in 1527, the son of one of the attendants in Henry VIII's court. His parentage was of lowly status but could boast important connections and Dee determined to overcome the former and make the most of the latter. His prodigious intellectual efforts (as part of which he seldom allowed himself more than four hours sleep) brought him a fellowship at the royal foundation of Trinity College, Cambridge. Dee was a showman who craved attention. That is not to deny his ability. He qualified in theology and law but his passion was for mathematics.

'By number, a way is had, to the searching out and understanding of everything able to be known.'* This statement, by Pico della Mirandola, lies at the heart of all Dee's research, writing and activity.

* Cf. F.A. Yates, *op cit.*, p. 80

He was captivated by numbers, which had a 'magic' of their own. And it was little short of miraculous that calculations done in the seclusion of a scholar's study could measure the size of the Earth, analyse the load-bearing potential of a structure or determine the aesthetic quality of a building.

But to understand mathematics was nothing less than to probe the mind of God, as the title of *De Divina Proportione*, a book published in Venice in 1509 by Luca Pacioli, indicated. This treatise (studied by Da Vinci, Piero della Francesca and other Renaissance artists) was just one of several on a subject that had intrigued philosophers since classical times and had featured in hermetic and cabbalistic writing.

The practical application of numbers was always important to Dee. After completing his basic studies in Cambridge, he spent some time in Louvain where he befriended Gerald Mercator, mathematician, geographer and instrument-maker, known to posterity as the father of modern cartography. The Flemish scholar had already freed map-making from the grip of Ptolemaic geography and produced several maps that changed the shapes of land masses and adjusted the measurement of terrestrial distances. He went on to perfect what would become known as 'Mercator's projection', which transformed the lives of mariners by enabling them to plot compass bearings on charts with longitude and latitude shown in straight lines. However, his work was not universally appreciated; he had spent several months in prison because the religious authorities suspected him of travelling abroad to consort with Lutherans. In 1550 Dee was briefly in Paris, lecturing on Euclid, before returning to England, where he was soon involved in the politico-religious issues dominating life at the Tudor courts.

It is difficult to grasp the breadth of this polymath's interests. The library he amassed in his home at Mortlake, to the west of London, was the largest private library in England and contained the writings of the Italian neoplatonists, numerous books and manuscripts rescued from dissolved monasteries and the works of his contemporaries. He was a devotee of Agrippa's occult philosophy and his laboratory housed 'scrying glasses', used in attempts to conjure up spirits, as well as alchemical apparatus. Amid his attempts to consult angels and to probe the innermost secrets of the universe, Dee directed his energies towards the practical needs of his contemporaries. He drew up navigational charts and wrote treatises for builders on the harmony of structures.

It is not surprising that a scholar of such an independent mind and forceful personality should have impressed the movers and shakers of his own and other lands. Courtiers, politicians and even heads of state were among his 'disciples'. In fact, he considered it his mission to play a major role in the destiny of nations. Hellenistic writers had urged rulers to keep philosophers at their courts to offer wise guidance, and Renaissance princes who considered themselves to be avant garde surrounded themselves with scholarly gurus.

Thanks to his standing in academic circles, Dee was welcomed in English court circles and attained the patronage of men close to the throne. In the reign of Catholic Mary Tudor he even survived examination for heresy and treason (attempting to kill the queen by poison or magic). But it was the accession of Elizabeth I that established his fame and influence. The queen and several members of her entourage were impressed by his philosophical teaching and his alchemical experiments. Dee now ventured into the realm of political philosophy. His *General and rare Memorials pertaining to the perfect art of navigation* (1577) embraced not only details of mathematical calculations valuable to mariners, but also enthusiastic advocacy of imperial expansion.

Dee enjoyed high favour for several years, being regarded as an official philosopher – almost an official prophet – of the Elizabethan regime. He prepared the queen's horoscope and was consulted on the propitious times for government initiatives, particularly with regard to voyages of exploration and settlement in the New World. Such political activities won him recognition by the elite but among ordinary people he had an unsavoury reputation as a sorcerer and devil-worshipper, a reputation that increased as he devoted more of his attention to conjuration of spirits.

His assistant in this activity was Edward Kelly, a self-professed medium. Opinion remains divided on whether Kelly (who had already suffered ear-cropping, the penalty for fraud) was a charlatan who took advantage of Dee's gullibility or whether he really believed in his occult powers. The question is academic; the important fact is that Dee believed completely that his 'scryer' could summon angelic beings. He believed because his intellectual/spiritual ambition was enormous. He sought to scale the heights of gnosis and discover those divine secrets which would unite all religions in the worship of the Christian God. It was always his contention that the Almighty had stirred within him 'an insatiable zeal and desire to know his truth'.

When his influence waned in England he travelled to foreign courts in search of patrons who would endorse his philosophy. The years 1583–1589 were spent in Bohemia and Poland, where the Emperor Rudolph II was among those who extended their patronage.

Not all Dee's Mortlake neighbours shared the admiration enjoyed by the celebrated magician in their midst. Scarcely had he departed when a mob, motivated by suspicious fear or greed or both, broke into his house to smash or remove the tools of his 'dubious' craft. Things did not go well on the Continent either. Kelly used his colleague as a stepping stone to his own advancement as an alchemist. Dee did not finally see through him until Kelly insisted – as a result of angelic revelation – on an experiment in wife-swapping. Dee's final years were unhappy. Though he was sometimes employed as an expert witness in witchcraft trials, he was largely neglected and reduced to near penury. He had become yesterday's man.

Tycho Brahe (1546–1601) was another scholar feared and hated by his fellow men. When vandals attacked his house and astronomical observatory in 1597, they were not content to trash the place; they razed it to the ground. That was understandable as Tycho was not a very pleasant man – he was an argumentative, self-willed braggart. In his youth he was famously involved in a dual that resulted in part of his nose being sliced off. Ever afterwards he drew attention to this 'badge of honour' by covering the wound with a metallic sheath. The Danish aristocrat also flouted convention by marrying a peasant girl. Like Dee, he was an exhibitionist and an eccentric who loved to impress people with his arcane knowledge. But unlike Dee he possessed considerable wealth with which to flaunt his bizarre behaviour. Guests at the Dane's lavish parties might be entertained by Jepp, a dwarf who doubled as jester and psychic medium, or they might be amused by the pet beer-swilling elk Tycho kept indoors. But this boorish, self-advertising nobleman was no empty-headed chinless wonder. Far from it. He was intelligent and independently minded, so much so that he abandoned the preoccupations of his class – law, war and dynastic alliances – for the study of the heavens.

Because educational resources in the Baltic states were poor and because Tycho's inquisitive mind could not be satisfied by the conventional courses on offer at Copenhagen University, he travelled to the Lutheran learning centres at Wittenberg, Leipzig and Rostock. It was while at Rostock in 1572 that a new bright star in the constellation of Cassiopeia became visible from Earth (what we now call a 'supernova'). Tycho was fascinated by it and intrigued to discover how prevailing Aristotelian science could offer an adequate explanation of this phenomenon. According to the established fixed-state theory, all objects beyond the moon's orbit were eternal and immutable, which would mean that the 'star' was not a star but something within the Earth's atmosphere. The way to test this theory was to observe the movements of the object. Closer bodies appear to move faster than those more distant. This one did not. The conundrum set Tycho on his life's work, the attempt to measure with ever greater accuracy the movement of heavenly bodies.

It was at this point that Tycho benefitted from a debt of honour recognised by Frederick II of Denmark. Back in 1565 the king had been saved from drowning by the quick action of Tycho's stepfather (who contracted pneumonia and died as a result). Frederick's gratitude now showed itself in generous support for his saviour's son. He appointed Tycho his personal astrologer and gave him the island of Hven, between Denmark and Sweden, together with generous funds for the establishment of a research centre, Uraniborg, dedicated to Urania, the muse of astronomy. This impressive establishment, the first of its kind in Europe, included a large mansion for Tycho and his family, an observatory, a workshop for the manufacture of astronomical instruments, an extensive library and an alchemical laboratory, as well as a printworks and a paper mill so that Tycho could publish all his findings.

This eccentric obsessive had hundreds of people working for him on the Uraniborg estate and he drove them very hard. To what end? Tycho Brahe was a plodder. With the aid of his army of virtual slave-labourers and working without the help of telescopes, he made a phenomenal number of observations and calculations about the movements of several astral bodies. He was forever developing and perfecting instruments such as quadrants and sextants for measuring the angles between Earth and the stars. He acknowledged the contribution made by Copernicus to the understanding of the workings of the universe, yet, this bold flouter of social convention was unable or unwilling to cast off the assumptions of Aristotelian physics, Ptolomanian cosmography and

Lutheran Biblicism. For example, he refused to accept that if a star was as remote as the more adventurous scholars suggested, it must be unimaginably large.

Tycho believed in a geocentric universe because all the weight of traditional astronomy was behind it and because Protestant and Catholic theologians were united in asserting that this was what the Bible said. His concern was not to overthrow conventional practice but to support it with more accurate data. This was vital, he asserted, for astrological predictions. Horoscopes were so important that they had to be based on the most accurate information available about stellar motions. The 'solution' he proposed to the conflict between received wisdom and the Copernican challenge was a compromise – a geo-heliocentric system. He held to the view that sun, moon and stars circled the Earth but claimed that the five planets of the solar system revolved around the sun. This, seemingly, satisfied the conservative establishment. When the works of Copernicus and Galileo (see p. 136) were placed on the Catholic index of banned books, Tycho's writings avoided that censure.

By 1597 he was a major Danish celebrity, internationally famous, hugely wealthy and living in great style. But he was no longer *persona grata* at the royal court. In fact, Christian IV, who had succeeded his father, Frederick II, conceived a real hatred for the scholar-aristocrat, whom he suspected of having an affair with his mother. (Might this have sparked in Shakespeare's mind the plot of another Danish prince struggling with the problem of his mother's infidelity, which had its first performance three or four years later?) Tycho was obliged to leave Hven, which was immediately set upon by vandals, perhaps at the instigation of the angry king.

But the Danish scholar did not want for powerful friends. Over the next couple of years he made a tour of the centres of learning in northern Germany, always accompanied by his enormous, impressive entourage and ended up in Prague at the court of the patron extraordinaire, the Emperor Rudolph II.

Rudolph was an obsessive collector of works of art and the latest technological inventions. John Dee had spent some time under the imperial roof and Edward Kelly had recently died while enjoying the emperor's 'hospitality' – the English mountebank had pushed his luck too far by promising Rudolph that he could make gold. The emperor's response was to shut Kelly up in a castle on the Bílina river near Prague and order him to prove his boast. Kelly died from injuries sustained in an escape bid.

Such salutary experiences did not, apparently, dampen Rudolph's ardour for supporting the most avant-garde thinkers of the day and he was delighted to have 'captured' Tycho Brahe from a Danish king who did not appreciate his subject's genius. He set his visitor up in another of his castles where the Dane continued his work and became the doyen of the scholarly community at the imperial court. Unfortunately, Tycho did not long enjoy his enhanced celebrity. He died of a urinary complaint in the autumn of 1601. It was, perhaps, inevitable that poison should be suspected and the rumour spread that Tycho had been murdered on the orders of King Christian. Centuries later, his 2010 exhumation enabled this suspicion to be discarded.

Twenty months earlier the third late Renaissance scholar we are considering here had also died a painful, though mercifully swift, death. Giordano Bruno (1548–1600) was a Dominican friar at Naples when he began his public career. He was ordained priest in 1572 but soon fell under suspicion for his freethinking tendencies. Exactly what he did believe in his formative years is not clear. His opposition to religious images suggests sympathy with Protestantism but he was an eclectic reader of banned books who went much further than others in his rejection of fundamental Christian dogma. Critics trying to get a 'handle' on Bruno's teaching accused him, among other heresies, of denying the divinity of Christ, pantheism and proposing the existence of innumerable parallel universes.

Bruno's tragedy was played out against the backcloth of conflict between Reformation and Counter-Reformation forces, which had by now reached steam heat. The Catholic Church and the major Protestant denominations all claimed to be the sole interpreters of Holy Writ. That meant that astronomers and magi fell under suspicion from both sides if they appeared to be questioning the plain truth of Scripture, and both sides were eager to display their orthodoxy by taking action against heretics. Christian IV of Denmark, the nemesis of Tycho Brahe, was particularly zealous in identifying and prosecuting witches. It was through the marriage of James VI of Scotland (later James I of England) to Anne, Christian's sister, that James also became an ardent persecutor of black magicians. In 1597 he gave the world the benefit of his thoughts on the subject in a slim volume entitled, *Daemonologie*. Where kings led, preachers and rabble-rousers followed, playing on the fears ordinary people entertained about occult knowledge and practice.

It is not surprising that the Faust legend revived with fresh vigour at the end of the century. Legends of the scholar who dabbled in forbidden knowledge had, for several decades, been part of common European folklore but now the idea of a scholar who sold his soul to the devil became part of the story. In 1587 an anonymous book, *Historia von D. Johann Fausten*, was published in Frankfurt. It was an immediate success and was translated into other languages. Within a decade Christopher Marlowe had written *The Tragical History of the Life and Death of Dr Faustus*, based on an English version of the German book. The play begins with a long speech in which the eponymous principal character deliberately turns his back on philosophy, medicine, law and theology in order to devote himself to that branch of *scientia* offered by the black arts.

> These metaphysics of magicians
> And necromantic books are heavenly;
> Lines, circles, letters, and characters:
> Ay, these are those that Faustus most desires.
> O, what a world of profit and delight,
> Of power, of honour, of omnipotence,
> Is promised to the studious artisan.

This was the conflict between good and bad *scientia* that Cornelius Agrippa had highlighted fifty years earlier. Marlowe's Faust deliberately turns his back on God and, with the aid of arcane spells, summons the demon Mephistopheles (another novel element that had not been part of the earlier legends), who assumes the shape of a Franciscan friar – observing that 'this holy shape becomes a devil best' (a sideswipe at the Catholic Church that would have gone down well with the Protestant groundlings in the Elizabethan theatre). Before the fatal contract is signed the scholar is visited by two angels, representing the duality of his nature. He rejects the injunction to think on heavenly things, dismissing them as 'illusions ... that make men foolish' and pursues what he convinces himself is a superior wisdom, promising 'honour and wealth'. What follows is a compound of high tragedy and low burlesque as Faust wanders the world, performing his magic to the entertainment of his audience, before paying the terrible price for his presumption. The dramatist presents a picture of the scholar whose hubris and inquisitiveness know no bounds. Nothing will deter him from pursuing the knowledge and experience of good and evil for which, according to the Bible, Adam and Eve were cast out of Eden.

Almost two decades later Shakespeare addressed the same themes in his last great tragi-comedy, *The Tempest*. He, too, has a magician, Prospero, for his central character. This weaver of spells also has a spirit to do his bidding but Ariel is no demon and the power he wields is white magic, not black. Prospero is the antithesis of Marlowe's character and more profound and sophisticated. He is the noble hermetist of the kind envisaged by Agrippa, using his powers only for good. Having brought his enemies to his island by means of his spells, he refrains from wreaking vengeance upon them. Rather than cling to his ethereal slave, he grants Ariel his freedom. Faustus becomes the victim of his baser nature but Prospero is at all times in command of himself and his destiny. By the end of the play all is restored, all forgiven and the magician forswears his magic: 'My charms are all o'erthrown and what strength I have's mine own,' he informs us. It is humanity that triumphs because Prospero has the strength of will not only to take up magic but to lay it down.

The difference between the two plays comes down to this: *Dr Faustus* is based on fable, while *The Tempest* finds its examples in real practitioners of mystic arts. While Marlowe was content to explore the theatrical possibilities of a popular tale, Shakespeare reflected the late Renaissance mood for probing and questioning strange phenomena. Contemporaries would have recognised in the fictional magi real scholars of recent memory who had pushed the bounds of human knowledge beyond what the moral guardians of the time considered permissible. In this time when old truths were being questioned and new truths sought after, many seekers achieved wide celebrity.

It was in the years between the writing of *Dr Faustus* and *The Tempest* that John Dee, Tycho Brahe and Giordano Bruno died and their stories were 'headline news' throughout Europe. We must now take up again the tragical history of the life and death of Fra Giordano Bruno.

Further wanderings had taken the renegade Dominican to Paris, Oxford and London, his celebrity increasing wherever he went. By the mid-1580s he had become quite a showman, gaining attention not only by his controversial (sometimes slanderous) writings and lectures but also by a remarkable gift of memory. He developed a system of mnemonics that enabled him to recall events and written texts with a precision which never failed to impress his hearers. In the French and English capitals he moved in court circles and enjoyed the patronage of Henry III, Elizabeth I and leading lights in the European cultural sphere. It is possible that he was also employed in intelligence gathering by Elizabeth's spymasters (an activity in which John Dee was also thought to be involved).

By 1585 Bruno was on the move again – to Germany and Poland, where he enjoyed the patronage of none other than that collector of magi, Emperor Rudolph II. Then, in 1591, he made the fatal mistake of returning to Italy. Two years later he found himself in a Roman jail where he suffered repeated interrogations and where, in 1600, he was burned as a heretic.

Believing firmly in his own superior wisdom, Bruno had confronted the religious and academic worlds of his day with intellectual arrogance born of frustration and anger. He was scornful of the beliefs and convictions that drove men into warring camps – Catholic, Protestant, Aristotelian, Neoplatonist. He was impatient with national rivalries and seems to have seen himself as something of a wandering ambassador intent on fostering peace between France and England. But, while he identified himself as a champion of peace and harmony, he displayed in his writings and his public disputations a cantankerous and disrespectful attitude towards his opponents.

Bruno's remarkably voluminous printed works, in the form of dialogues, poems and plays, covered a range of religious, philosophical, ethical, political and scientific subjects. It is difficult for the modern reader to understand his cosmological system. Bruno's biographer, Frances Yates, observed that he had 'an extremely strange religion'. It was hard to know whether to refer to 'his philosophical religion or his religious philosophy or his philosophical-religious magic'.* In all probability, there was no 'system', just a series of reactions to the warring religious and philosophical arguments that were tearing European Christendom apart. He stood in the tradition of Cornelius Agrippa (though even the word 'tradition' is too tidy in this context) in searching for an overarching philosophy, a new (and yet ancient) corpus of belief that might unite all men of faith and goodwill.

Bruno was among the few contemporary thinkers to embrace the Copernican heliocentric theory. During his stay in England (1583–1585) he enjoyed the patronage of some of the leading intellectuals and trendsetters of the Elizabethan court. This opened up the possibility of delivering a series of lectures in Oxford in which he expounded Copernican cosmology. Unsurprisingly, the ideas expressed by this excitable foreigner did not go down well. George Abbot, the future Archbishop of Canterbury, expressed the opinion of many (not just churchmen) when he reported, 'That Italian [clown] . . . undertook . . . to set on foot the opinion of Copernicus that the Earth did go round and the heavens did stand still, whereas in truth it was his own head which rather did run around and his brains did not stand still.'† Some historians of science have taken this as an example of a closed religious mind in the face of scientific truth. That misses the point. Bruno was not defending pure, scientific enquiry *against* religion. On the contrary, the attraction of heliocentric theory (which he imperfectly understood) was that it could be made to fit into his more profound (as he thought) hermetic schema. In *The Ash Wednesday Supper*, a dialogue published during his stay in England, he had this to say about Copernicus:

* F. Yates, *Giordano Bruno and the Hermetic Tradition*, University of Chicago Press, Chicago, 1964, pp. 260–2

† Cf. R. McNulty, 'Bruno in Oxford', *Renaissance News*, Vol. 13, No. 4, Winter 1960, p. 304

> He was possessed of a grave, elaborate, careful and mature mind ... a man who in regard to natural judgement was far superior to Ptolemy, Hipparchus, Eudoxus and all the others who walked in the footsteps of these; a man who had to liberate himself from some false presuppositions of the common and commonly accepted philosophy or perhaps I should say blindness. But for all that he did not move too much beyond them, being more intent on the study of mathematics than of nature. He was not able to go deep enough ... and set attention firmly on things constant and certain.*

Bruno championed Copernican heliocentricity because it appeared to prove the truth of that pseudo-hermetic system of a living, organic cosmos sustained by an omnipresent deity that pre-dated the limited vision of Christianity and other world faiths. Science was, for him, a tool for building a better world but that world was based firmly on religion.

Unfortunately for him, that religion was not one that could be recognised as orthodox by any of the major Christian churches. By itself, Bruno's adventurous cosmography would probably not have proved fatal to him. In England scholars were prepared to listen, albeit sceptically, to his views. On the other side of the Channel, much of northern Europe had embraced Lutheranism by the end of the sixteenth century, and within Lutheranism there existed a humanist tradition that took inspiration from Luther's assistant at Wittenberg, Philip Melanchthon (and was, in fact, known as 'Philippianism'). The spirit of honest and devout enquiry pervaded the universities of North Germany and Scandinavia. Fresh thinkers such as John Dee and Tycho Brahe could pursue their researches

* *The Ash Wednesday Supper*, trs. S. L. Jaki, First Dialogue, University of Chicago Press, Chicago, 1969

and even attract valuable patronage. The suggestions that man's terrestrial home was not the centre of the universe, that that universe was infinite and that other inhabited worlds might exist – these were shocking but were not burning matters. However, denying the divinity of Christ, endorsing the conjuration of spirits, rejecting the doctrine of transubstantiation – such were the issues that brought an unrepentant Bruno to the stake.

The Renaissance and the Reformation had set the intellectual world adrift from its ancient moorings. Thinkers were venturing beyond the horizons of accepted truth, just as mariners were extending their reach across the oceans. All were intent on learning more of God's wonders written in the book of nature. The scholar and the sailor both followed lonely and potentially dangerous occupations. They might achieve fame. They might also experience rejection, as John Dee, Tycho Brahe and Giordano Bruno, each in their different ways, discovered. And always, in the dark recesses of their minds there lurked the warning of Dr Faustus' terrible, eternal fate.

CHAPTER FIVE

Medicine men

Two approaches to philosophy that diverged with increasing clarity during the Renaissance were the Aristotelian method based on the drawing of conclusions from accepted first principles and the creation of principles from observation and experimentation – eventually known as the inductive method (see p. 115). For professional healers there was always a third, and more important, consideration: treatment – what we now call 'clinical medicine'. Perhaps it would be more accurate to say that medical practice combined traditional theory and experimentation.

The doctrine of the four humours reached back, via Aristotle, to the fifth century BC. Hippocrates still dominated medical theory and practice but hands-on care of patients involved the trial-and-error applications of herbal remedies, surgery, purging, blood-letting, diet control, temperature adjustment and other potentially curative methods. To the skilful doctor every surgical operation and every medicinal prescription was an experiment whose results were carefully monitored and recorded.

The works of Galen (Aelius Claudius Galenus, *c.* AD 130–*c.* 200) dominated theory and practice well into the modern era and were certainly based on inherited theory and induction. This philosopher-doctor spent most of his professional life tending the gladiators in the Roman games. He was one in a long line of practitioners who were kept 'on their toes' by treating the varied injuries sustained by men in combat. Not only did his practice include most known aspects of medicine and surgery, but he also left a monumental library of written works. Several were passed down in the West over the centuries and others arrived from *c.* 1000 onwards via Byzantium and the Islamic world. So comprehensive was Galen's contribution towards the study of the body and its well-being that we should not be surprised that generations of teachers and students regarded the Galenic corpus as the Bible of the profession, contradiction of which was a form of heresy.

By the sixteenth century the care of the sick was in the hands of a wide variety of practitioners. Most people could not afford professional care and had to rely on the ministrations of local 'wise' men and women who brought to their patients a mixture of herbal remedies, spells and potions. Beyond these homespun healers lay 'experts' claiming a greater degree of skill and knowledge and eager to cash in on physical suffering. Apothecaries were to be found on every high street and in every market, doing their extravagant best to impress the public.

> ... in his needy shop a tortoise hung,
> An alligator stuff'd and other skins
> Of ill-shaped fishes ... and old cakes of roses were thinly
> scattered to make up a show ...
>
> William Shakespeare, *Romeo and Juliet,* Act V, Scene I

Above the world of amateur potion sellers and fraudsters stood the serious practitioners of medical science. Many of them belonged to monastic institutions. All larger religious houses had their infirmarians whose primary responsibility was the health of their colleagues and care for the local poor and needy, which had always been part of the monastic vocation. From this responsibility sprang the first European hospitals. As well as clinical care, these establishments provided hospitality to travellers, refuge for the homeless and doles of food for the destitute. Some became major institutions in their own right. The Hôtel-Dieu, the oldest such hospital in Europe, was built in the shadow of the Louvre, in Paris, in the seventh century.

Five hundred years later, the monk and courtier, Rahere, founded a priory and hospital in London dedicated to St Bartholomew. 'St Bart's', England's oldest hospital, survives to this day. Such larger and more prestigious foundations attracted university-trained doctors as visiting physicians.

Most monks and nuns worked on a much more modest basis. They were not famous for groundbreaking research or medical innovation. There were exceptions, however. The twelfth-century Hildegard, Abbess of Bingen, wrote two important treatises, *Physica* and *Causae et Curae*, in which she described the workings of the human body, the curative properties of various natural elements and the technique appropriate for their applications. Many religious houses became repositories of medical knowledge, written down for the benefit of later generations of infirmarians. Works known as herbals described the cultivation and use of plants believed to be efficacious. Such books aided diagnosis by listing the symptoms of diseases and ailments.

The approach to curative medicine in all these religious hospitals was holistic. Like the Greeks before them, Christian healers thought of the human body not as a mechanism to be kept in efficient working order, but as the temporary dwelling place of an eternal soul. For over a thousand years, inhabitants of the classical world had travelled to temples called asclepieia, dedicated to the god Asclepius, seeking the healing of mind and body. They received the ministrations of priests skilled in the treatment of physical ailments and in what we would now call psychosomatic disorders. Through medicine, religious rituals, music and contemplation of works of art, patients were helping to align themselves with the divine order of the universe. According to many inscriptions that have been discovered at these sites, miraculous healings were not uncommon.

The concept of health through harmony with the cosmos is reminiscent of St Paul's recipe for spiritual well-being: 'Fill your minds with those things that are good and deserve praise; things that are true, noble, right, pure, lovely and honourable' (Philippians, 3:8). The regimen of the medieval monastic hospital was, thus, not new or 'unscientific'. It offered medical treatment and prayer in an atmosphere of regular worship, religious music and reflection on eternal verities.

The Reformation changed all this. Throughout much of northern Europe, religious houses were either closed or allowed to expire slowly as their inmates died or were encouraged to leave. The impact on everyday clinical medicine was varied. Some hospitals disappeared. Others were taken over by civic authorities. St Bart's, for example, was refounded by Henry VIII and entrusted to the care of the City of London corporation. Its new official name was the House of the Poor in West Smithfield. Some hospitals were 'secularised' for other reasons. The Hôtel-Dieu in Paris found itself in financial difficulties early in the sixteenth century and was taken over by a committee drawn from the mercantile community. In many places the gap left by monks and nuns was filled by lay benefactors. Wealthy donors, no longer needing to leave legacies for the performance of masses, made provision instead for the foundation and maintenance of hospitals and almshouses.

The door was now wide open to medicine men who came from different backgrounds. They fell into two categories, frequently at fierce rivalry with each other. Most physicians were university graduates (Padua and Heidelberg currently possessed the best schools of medicine) and had studied the old masters, among whom Galen still held pre-eminence. They considered themselves the only fully qualified practitioners, having studied astrology and plant remedies as well as anatomy.

Knowledge of anatomy was slowly becoming easier to obtain as the old prejudices about dissecting dead bodies gradually eased. However, there were still tales of sordid deals done with prison authorities over the remains of executed criminals and hair-raising anecdotes about grave robbers. The other 'experts' were the barber-surgeons whose education was gained at the sharp end, in hospitals or private practice, consisting largely of the treatment of minor injuries. The advance of medical science was undoubtedly hindered by the rivalry between these two groups of specialists, who were jealous of their reputations and guarded professional secrets that would have been far better shared. But the new opportunities for practice produced some notable individuals who, in various ways, influenced the development of science.

Ambroise Paré (1510–1590), a boy of nine in the year that Leonardo da Vinci died, came from humble origins and there was no question of his receiving a university education. He chose the career of barber-surgeon and learned on the job at the Hôtel-Dieu. Paré was brought up in the atmosphere of hands-on science and had no time for theory. In treating patients, the only thing that mattered to him was discovering what worked. He rose rapidly in his profession and became surgeon to the French royal court. Although he could have enjoyed a comfortable life, attending the rich and famous and performing operations at the Hôtel-Dieu, where he was appointed to the surgical staff, he chose instead to spend much of his time on military campaigns.

France's frequent wars, as well as creating a need for field doctors, provided abundant opportunities for Paré to experiment with various methods of healing wounds and mending broken limbs. He later wrote an entertaining memoir in which he boasted of his trial-and-error methods. Describing an incident when he was tending soldiers suffering from gunpowder burns, he explained that in his early years he dressed such wounds 'by the book' with scalding hot oil. On occasions when that ran out he applied instead 'a digestive of eggs, oil of roses and turpentine' and was delighted to discover that this ad hoc treatment worked far better. Later, having met a surgeon famed for his expertise with such wounds, he laboured for two years to learn the secret of his success.

> In the end, thanks to my gifts and presents, he gave it me, which was to boil in oil of lilies, young whelps just born and earthworms prepared with . . . turpentine. Then I was joyful and my heart made glad, that I had understood his remedy, which was like that which I had obtained by chance. See how I learned to treat gunshot wounds; not by books.*

Paré was an empiricist, but not in the sense of a scientist shutting himself in a laboratory to conduct minute observation of causes and effects. He worked in the manner of a practitioner, trying out remedies on living (and dying) people. Nowadays we have laws about that sort of activity but five hundred years ago it was a kind of experimentation that sometimes produced beneficial results. His trial-and-error approach left an impressive legacy to future generations of surgeons. He championed ligature of arteries after amputation, rather than cauterisation (the shock of which caused more deaths than the operation). He made the first tentative steps towards anaesthetics by using laudanum to relieve pain during surgery. Several pregnant women had cause to be grateful to Paré and the surgical innovations he pioneered. He worked on the safe delivery of breech-birth babies, which previously had only been extracted by being cut in pieces. He is regarded as the founder of modern forensic medicine because he set out guidelines for the way medical evidence should be presented in court. He invented his own instruments for performing procedures. Not least among his accomplishments was his publication of his discoveries, even though contemporary purists disparaged his books because they were written in French

* S. Paget, *Ambroise Paré and his Times*, G. P. Putnam's Sons, New York, 1897, p. 35

and not the Latin of scholars. Pomposity, professional jealousy and rivalry between physicians, surgeons and apothecaries continued to hamper the development of medical science.

> When . . . the whole compounding of drugs was handed over to the apothecaries then doctors lost the knowledge of simple medicine which is absolutely essential to them . . . when the doctors supposed that . . . mere knowledge of the viscera was more than enough for them, they neglected the structure of the bones and muscles, as well as of the nerves, veins and arteries . . . when the whole conduct of manual operations was entrusted to barbers, not only did the doctors lose the true knowledge of the viscera, but the practice of dissection soon died out.*

That analysis of Renaissance medical studies was from the pen of the man often referred to as the father of modern anatomy – Andreas Vesalius (1514–1564). He came from a long line of medical practitioners living in what is now Belgium and was then part of the Habsburg Empire and enjoyed a glittering career. After completing a foundation course at Louvain University, Vesalius moved to Paris to study medicine. There, despite being an apt and zealous student, whose extra-curricular activities included stealing bodies to practise on, his progress was interrupted by war between France and the Empire. A period of further study in Italy ended with him being appointed Professor of Surgery and Anatomy at Padua in 1537. He was an inspiring lecturer and his classroom was regularly packed with eager young men, drawn by his style and his lack of reverence for Galen and the other ancient masters whose theories were not

* Andreas Vesalius, *De Fabrica Corporis Humani* (1543), preface in *Proceedings of the Royal Society of Medicine*, July 1932, p. 1360

standing up to the new methods of hands-on investigation of human cadavers.

As erroneous theories based on the assumption that most mammals shared a similar anatomy were progressively abandoned, there was need for a new 'geography' of the human body. In 1537, aged only twenty-three, Vesalius made his first foray into scientific publication with a study of the vein system. In 1543 he produced his master work, *De Humani Corporis Fabrica* (*On the Workings of the Human Body*). The text was superbly illustrated with plates from the workshop of the Venetian artist Titian, and Vesalius, having in mind the needs of impoverished students, issued a shorter and cheaper version of his book in which the images predominated.

This second edition was a striking example of the impact of printing on the spread of knowledge. Now it was not only established scholars and wealthy courtiers who had access to the latest developments in intellectual fashion; the next generation were finding it easier to keep themselves informed. In 1546 Vesalius ventured beyond the study of anatomy with *Epistle of the China Root*, which contained further criticisms of traditional treatments – in this case casting doubt on certain herbal remedies. This marked the end of an astonishingly brief period of mould-breaking literary output.

Vesalius' work led to a huge leap forward in the understanding of how the human body worked. He established accurately its skeletal framework and showed how muscles provided the means by which the parts of that framework moved in relation to each other. He freed knowledge of the nervous system from the Aristotelian myth that the heart was the centre of all physical sensation and emotion by pioneering the study of nerves and their emanation from the brain. With accurate dissection of the heart he opened up the way for later scholars to discover how blood circulates. He laid the foundation for understanding the parts of the digestive system and their functions.

Inevitably, there were experts who did not like traditional theories being challenged. Attacks by older and 'wiser' men may have contributed to Vesalius' decision to abandon teaching and take up the lucrative post of physician to Emperor Charles V and, subsequently, to his son, Philip II of Spain. As well as treating members of the court, his new position involved accompanying imperial armies on campaign. It seems very likely that Vesalius and Paré must sometimes have found themselves treating war casualties on opposing sides.

However valuable Vesalius was to his employer, he could not escape criticism from medical men of the old school. Spain was a dangerous place to make enemies for it was there that the Inquisition wielded enormous power. Ever vigilant in their determination to weed out theological novelties, the agents of this unyielding institution were feared by people at all levels of society. At some stage suspicion fell upon the beliefs and practices of Vesalius which, it was suggested, contradicted the teachings of the Church. The details of the case are obscure. It seems that, after examination, he was exonerated. But this did not stop his rivals pursuing their whispering campaign.

Another contemporary was not at all reticent about publishing his opinions on a variety of subjects despite the fiercest opposition from many quarters. Michael Servetus (*c.* 1509–1553) was some four or five years older than Vesalius. There is much about his life that is a mystery. He was probably born in Villanueva de Sijena in north-eastern Spain. When he needed an alias (as he frequently did) he sometimes chose to call himself Michel de Villeneuve. His birth date could have been 1509 or 1511. What can be said with reasonable certainty is that he was born into a respectable, orthodox Catholic family. One of his brothers was a priest and Michael was destined for a career in the law. He studied at Toulouse but after a few years left that to take a position in the retinue of Emperor Charles V. Servetus had a brilliant and enquiring mind and a young man's conviction of the superiority of his own intellect. Critical of the beliefs and practices of contemporary religious and political leaders, he did not hesitate to publish his opinions for the benefit of a wider public.

We next find him, in his mid-twenties, studying medicine in Paris, where he sometimes found himself working at the dissecting bench beside Vesalius. Like his brilliant colleague, he questioned some of the traditional teaching of his professors. However, one aspect of the physician's craft he firmly believed in was the practice of astrology. He rushed into print with his ideas on medicine and delivered lectures on mathematics and astrology. This won him no admirers among his academic superiors and Servetus soon found himself in hot water, accused of basing his teaching about divination on the pagan practices advocated by Cicero. He had to move to Montpellier to complete his studies and it was there that he set himself up as a medical practitioner.

For the next few years (1539–1547) Servetus expended his mental energy on a wide variety of subjects. Centred in Lyon and Vienne, he worked for publishers as well as following his medical profession. He wrote treatises or edited the works of others on anatomy, pharmacology, geography, astrology but, above all, theology, to which, in his view, all sciences were subject. Should we regard him as a polymath or a dilettante? Neither appellation seems entirely appropriate. If by polymath we mean someone profoundly knowledgeable in several areas of study, the word does not apply. Much of what he wrote was in hasty response to the criticisms of other scholars rather than in persistent pursuit of solutions to eternal mysteries. Yet the passion he brought to his studies prevents us dismissing him as a mere dabbler. When thinking of Renaissance scholarship we have to remember that *scientia* was a unity. All knowledge emanated from and found its consummation in God. It did not seem strange to intellectuals to approach knowledge of the Creator and his purposes along different roads simultaneously. Rigid specialisation still lay in the future.

Few Renaissance scholars brought a more fiercely independent spirit to everything they studied than did Servetus. His reading encompassed Greek and Hebrew documents, Jewish Old Testament commentaries, hermetic writings, second- and third-century Gnostic texts, medieval Neoplatonic treatises and the works of Greek and Latin Church Fathers. From this amalgam there emerged a concept of the cosmos that was as original as it was individual.

To the history of medicine Servetus made a couple of significant contributions. He was the first person to correctly identify the relationship of the heart and the lungs. Galen had taught that the aeration of blood occurred in the left ventricle of the heart. Servetus observed that blood is 'transmitted from the pulmonary artery to the pulmonary vein' by passing through the lungs, where it is 'mixed with the inspired air and purged of fumes by expiration . . . it is not simply air, but air mixed with blood which is sent from the lungs to the heart . . . the bright hue is given to the [arterial] blood which is by the lungs, not by the heart'.* This may reasonably be hailed as a first step towards the discovery of the circulation of the blood. Yet it did not receive the acclaim that it should have done because it was not set forth in a medical treatise.

In a book entitled *Restitutio Christianismi,* Servetus attempted nothing less than a complete overhaul of Christian doctrine (see p. 109). For example, he sought to identify something about which scholars had long argued – the precise location of the human soul. Earlier theories had variously identified the heart, the liver or the brain as the place where the essence of man resided. Servetus' study of the Bible and Jewish commentators led him to Genesis 9:4, where God laid down for Noah and his descendants the rules about kosher food: 'The one thing you must not eat is meat with blood still in it; I

* Cf. R. H. Bainton, *Hunted Heretic – The Life and Death of Michael Servetus,* Beacon Press, Boston, 1960, p. 121

forbid this because the life is in the blood.' Servetus pounced on this and found that his anatomical studies provided proof of the text written down centuries before. Man's soul was, he asserted, a vibrant, living, dynamic thing. How could it be identified with any of the body's static organs? But the blood was a traveller, constantly on the move, purified in the lungs and transiting the brain where it was animated by God and conveyed life to all the body's functions. Had he followed this line of reasoning Servetus might have anticipated by more than eighty years William Harvey's monumental discovery. Unfortunately for his reputation in the medical community, he made his point in a notoriously controversial religious work and, since that was rejected by Church authorities, contemporary physicians declined to take it seriously.

Servetus' other original contribution to medicine took the form of a foray into pharmacology. *Of Syrups* was a discourse on sweet decoctions used for making tonics and herbal remedies palatable. However, in it the author did not confine himself to the relative efficacy of various ingredients: he was more interested in the process of digestion. This had been provided by the Creator as the means of segregating, assimilating and discarding the elements of the body's food intake. It followed that no medicine should be regarded as a 'cure' for physical disorder. Rather, its efficacy lay in assisting natural (i.e. divinely regulated) processes.

On this point the last in the quartet chosen to represent Renaissance medicine would have agreed with Servetus, though he embraced very few of the ideas of that man – or indeed of anybody else. Philippus Aureolus Bombastus von Hohenheim (1493–1551) was committed to nothing as fiercely as he was to controversy. Contemporaries regarded him either as an eccentric quack or a misunderstood genius – a difference of opinion that persisted for centuries after his death. An evident personality disorder exacerbated by a drink problem made him intolerant and intolerable, a maverick destined to be ostracised by the scholarly world but idolised by many then and since as a seer more insightful than ordinary mortals.

In strict chronological terms he was the first of our chosen quartet of medical men but he was such a stand-alone character and his activities covered so many fields of enquiry that we must differentiate him from the community of Renaissance doctors. Like many of them he was born into a medical family, his father being a German town physician. Like them he qualified at university (Ferrara). Like them he travelled widely in pursuit of knowledge. But there all similarity ends. His rejection of most aspects of traditional learning and practice was dramatically marked in 1527 when he made a public bonfire of standard medical works relied on by all contemporary practitioners.

To underline his contempt for other followers of his chosen profession, he set aside his family name and demanded to be known as Paracelsus. It was in 1478, only five years before his birth, that *De Medicina*, by the Roman writer Aulus Celsus, had been first printed and immediately welcomed as a valuable addition to the library of every erudite physician. By calling himself Paracelsus, 'Better Than Celsus', the truculent German left no doubt in the minds of all he met just what he thought of accepted wisdom: 'My shoebuckles are more learned than Galen', he told his growing body of outraged critics. That was only the most printable of the insults he hurled at everyone who presumed to disagree with him. Unsurprisingly, he seldom spent more than a few months in any town and was more often than not obliged to abandon his practice and his patients. Rejection by his peers drove him to express his ideas and pour forth his vitriol in numerous writings, most of which were not published until after his death.

He adhered firmly to the Renaissance concept of the unity of all *scientia*, and his numerous works embraced toxicology, theology, magic, cosmology and apocalypticism. Sifting out his contributions to medical knowledge from this welter of arcane literature takes us first to his *Opus Paramirum* (1530). In this he laid down as a basic principle that man is constituted of three ingredients – sulphur, mercury and salt. These were the chemical results of the interaction of the four basic elements accepted by all physicians – earth, air, fire and water. All disorders resulted from imbalance of these constituents and could be treated by application of one of the three principles or its derivatives. Thus, he advocated mercury as the best treatment for syphilis. To the study of medicinal remedies he bequeathed the observation, 'The dose makes the poison.' Many substances regarded as harmful, he asserted, might be efficacious in small amounts.

On the basis that man is essentially a bundle of chemicals, Paracelsus explored the animal, vegetable and mineral kingdoms in his quest for remedies. He experimented with iron, antimony and various mineral salts. He sought out 'similars'. Thus, ailments of the ear could be treated with distillations of cyclamen because the leaf of that plant resembles the human ear. When dealing with disorders in internal organs he prescribed medicines extracted from the corresponding parts of animals. Anatomy he dismissed as a useless occupation. What could possibly be gained from the study of dead bodies? he demanded scornfully. By eschewing book learning and employing novel – often bizarre – methods he made some remarkable discoveries. Silicosis, he asserted, was caused by the inhalation of metal dust and a goiter was the result of drinking mineral-contaminated water.

In *Die Große Wunderznei* (*The Great Surgical Treatise*) of 1536 he took further his theories of chemical cures. His approach to the art of healing was what we would now call 'holistic'. The good physician must treat the whole person, not just the disease or injury. Understanding the state of a patient's mind, he claimed, was fundamental to arriving at the appropriate treatment. He required the sick to record their dreams in as much detail as possible. This unique *Wundermann*, disowned by the medical profession as a charlatan but revered by a growing number of people impressed by his almost mystic language and outrageously novel practices, had anticipated by some four and a half centuries the discovery of psychosomatic disorder.

This was just one result of his probing the meaning of the cosmos and man's place in it. The art of healing, he claimed, demanded mastery of other disciplines. First came metaphysical philosophy. The physician must see himself as a microcosm of the universe and seek illumination from its Creator. By direct intuition he would gain the necessary knowledge to deal with the cases presented to him. Next in importance was astronomy. The human frame 'corresponded' to the solar system: it contained elements related directly to the planets. Detailed knowledge of their motions was, therefore, vital. The third discipline was alchemy, which Paracelsus defined as the 'chemistry of life', which must be understood before appropriate medicines could be prescribed and prepared. Above all the physician must be a man of virtue. Only the practitioner who was pure of mind and heart and not sullied by ambition, avarice, vanity, envy or conceit would ever be master of his craft.

Paracelsus, Vesalius, Paré, Servetus – four medical pioneers all active in the first half of the sixteenth century. The timing is highly significant. They were men of the Renaissance, freethinkers, who challenged old assumptions and dared to propose new ways of practising medicine. But they were also men of the Reformation, allied intellectually with the bold spirits who were standing up to the religious establishment. Just as some physicians rejected the outworn theories of Galen, so religious radicals were critical of scholasticism and papal decree. But these were not two distinct categories. As we have seen, *scientia* was not compartmentalised. Nor was life and in the 1500s religion was a fundamental part of life. Paracelsus was twenty-seven and the others were still children in 1517, when Martin Luther challenged the indulgence traffic and set a match to the fuse of incipient rebellion. But by the 1530s, when the chain reaction of explosions was devastating Europe, all four were in the thick of the conflict.

After his extraordinary, brief years of creative publishing and lecturing activity, Vesalius settled as physician to Charles V and Philip II, following the court around Spain and the Low Countries (which were under the control of the imperial crown). During the 1540s and 1550s the government was fighting an increasingly bitter battle against the spread of Lutheran and Calvinist 'heresy'. Rulers were egged on by the Inquisition, whose officials exercised enormous power and prided themselves on being no respecters of persons in their zeal to root out religious error. As we have seen, Vesalius was briefly caught in the inquisitorial net. His anatomical lessons were questioned and Charles V handed him over for interrogation. There can be little doubt that Vesalius faced a combined attack from religious leaders and reactionary members of his own profession. He escaped censure on that occasion but he had to walk warily, displaying whenever possible his loyalty to orthodox Catholicism. In 1564 he embarked on a pilgrimage to the Holy Land. While in Jerusalem he received an invitation to return as professor to Padua and he immediately set out for Italy. He was only fifty and a new phase of his career beckoned. Unfortunately, he died when his ship was wrecked off the island of Zakynthos. This gifted but impoverished medical pioneer was buried in an unmarked grave, the expenses being met by an unknown benefactor.

Ambroise Paré, by contrast, lived to be eighty. Yet he, too, had to negotiate the rocks and rapids of religious controversy. He spent the last three decades of his life in Paris as court physician to four Valois kings. The French capital was not a comfortable place to be. In 1559

King Henry II was injured in a tiltyard accident. His adversary's lance pierced his helmet and struck him above the right eye. Paré had to try to save the monarch's life. Vesalius was among a group of physicians who also came to help. They dissected the heads of dead criminals in their efforts to discover what damage might have been caused by the penetration. It was all to no avail; Henry died after ten days of excruciating pain. He left a widow, the notorious Catherine de Medici, and three underage sons.

The queen mother's attempts to maintain the authority of the Crown was bedevilled by rivalries between the leading noble houses and violent conflict between Catholics and Huguenots (French Protestants). According to some sources, Paré belonged to the latter and was obliged to keep his beliefs secret. The young king, Francis II, survived his father by only seventeen months. A rumour that Paré poisoned the lad on the orders of Catherine is probably more indicative of the atmosphere of fear and intrigue at court than of real filicide.

The royal doctor was in Paris in August 1572 when the appalling St Bartholomew's Day Massacre occurred. There is little doubt of the queen mother's complicity in the event that sparked off the atrocity: the attempted assassination of Gaspard de Coligny, one of the Huguenot leaders. Paré tended the stricken victim, who would probably have recovered had not his enemies broken into his bedchamber to finish the job. This murder was the signal for an onslaught on all the Protestants in the city. For days, bloodlust and fanaticism ran unchecked through the capital and spilled over into the surrounding country, claiming thousands of lives. Paré might well have ended up among the slain had he not been protected personally by King Charles IX. The royal physician lived as a covert heretic in this poisonous atmosphere throughout the French wars of religion and was present at the deathbeds of two more kings: Charles, who died of phthisis, and Henry III, who was assassinated.

While Vesalius and Paré were intent on avoiding religious controversy, Paracelsus and Servetus embraced it as part of their philosophy of life. Servetus was only twenty when he published his *Seven Books on the Errors of the Trinity* (1531). His motivation was a reading of the Bible in the original languages and in a spirit of free inquiry. He rejected scholastic doctrine as an attempt to square what he considered the clear teaching of Scripture with Greek philosophy. However, his own prospectus was drawn from a variety of earlier sources, including Neoplatonist writings, Gnosticism, rabbinic speculation and even Hermes Trismegistus. There is no doubting the depth and breadth of Servetus' scholarship.

Specifically, Servetus argued that the doctrine of the Trinity, which had been defined at the Council of Nicea in the fourth century, had no scriptural justification. He belonged to that amorphous Reformation fringe that went beyond the bounds of the mainstream Protestant churches and was anathema to them as well as to the Catholic hierarchy. They took the Bible as the only authoritative basis for Christian belief and interpreted it rationally. They also applied the litmus test of ethical empiricism. If Luther, Calvin and co. were right, why did their followers not live manifestly better lives than the adherents of the Pope? The individualism of these extremists produced a rainbow-hued 'left wing' of the Reformation. Various groups espoused such principles as pacifism, rejection of political and legal restraints, moral perfectionism, apoliticism and separation from the world. Servetus and those who thought like him were closest to Renaissance humanism in their rejection of medieval dogma and their open-mindedness to new insights.

Just as any questioning of Galen and other ancient authorities provoked the ire of the medical establishment, so rejecting the received wisdom of the medieval schoolmen brought bitter reaction from conservative churchmen. But it was not papal champions who entered the lists against Servetus. John Calvin, leader of the reform in Geneva, took an extremely personal dislike to this truculent radical. He had produced, in the *Institution of the Christian Religion*, his own systematic theology in answer to the scholastic schema and was determined to defend it against all comers. Anti-trinitarianism was, in his view, giving the well-reasoned Protestant position a bad name. He denounced the Spaniard's writings as heresy.

Servetus' response was twofold: he elaborated his theories further in print but he also retired from view behind a false identity. He devoted himself to medicine and proofreading. Yet he could not resist continuing the controversy with Calvin. Sheltering behind anonymity, he engaged in a written confrontation that became increasingly bad-tempered. Matters came to a head in 1553, after the publication of Servetus' comprehensive *The Restoration of Christianity*. This made Calvin so angry that he actually drew the attention of the Inquisition in Lyon to this dangerous heretic. Servetus was arrested, but escaped. Now, instead of slipping into safe obscurity, he made for Geneva. Perhaps he hoped to engage Calvin in public debate. Perhaps it never occurred to him that a Protestant state would send a man to his death for his beliefs. In October, he was arrested, condemned and burned at the stake, becoming the most famous (or notorious) martyr to suffer that fate in a non-Catholic state. What this demonstrated was that reformers such as Calvin, who championed freedom of belief and expression when they were in a powerless minority, could quickly change their priorities when they were in a position to enforce their understanding of truth.

The execution of Servetus outraged people across Europe, not a few of whom were Calvin's firm supporters. As one observed, 'To kill a man is not to defend a doctrine, but simply to kill a man.' Indeed, Servetus' doctrine was not consumed by the flames at Champel. His beliefs were fundamental to a new sect of the Christian clan – Unitarianism – which spread after his death, first of all to Poland, Hungary and Transylvania and eventually round the world.

Had Paracelsus lived beyond the mid-point of the century and into the era of the of Wars of Religion he might well have shared Servetus' fate. If we want to get to the root of his thinking (no easy task) we have to recognise that his main concern was religious. The conundrum he set himself to unravel was the relationship between God and man. His spirited opposition to the dogma promulgated by the Establishment extended beyond the leaders of the medical profession to the proclaimers of Catholic and Protestant truth. He had considerable sympathy for the 'heretical' freethinkers on the fringes of official Christianity but he belonged to no church, sect or protest group. He could never share the official dogmas of any community. He impishly dismissed the Pope and Luther as 'two whores debating chastity'.

It is easier to list the things Paracelsus did not believe than to formulate a 'Paracelsian theology'. When he was dealing with the mysteries at the heart of Christian faith he came closer to early Gnostic thought than to any other recognizable intellectual schema. Rather than humbly saying with the psalmist, 'Such knowledge is too wonderful for me; I cannot attain unto it' (Psalm 139:6), he was driven to propose – or, rather, to confidently assert – his own explanations. Thus, for example, his answer to the puzzle of the Holy Trinity was a kind of amoebic self-division within the Creator, who was, initially, one but willed himself to become three. At the Incarnation God did not 'take flesh of the Virgin', thus becoming completely human. Rather there was a merging in the womb that resulted in Jesus being born with 'holy flesh'. Paracelsus never doubted that divine mysteries were capable of intellectual elucidation and that he, Paracelsus, was uniquely gifted to provide that elucidation.

It is not surprising that such a sharp-edged character should have provoked opposition from many quarters, nor that he was seldom able to settle for very long in any one place. But any attempt to include Paracelsus among the ranks of political revolutionaries fails. In the mid-1520s, when he was at the height of his powers, central Europe was convulsed by the Peasants' War. Radical preachers thundered against a hierarchical society in which the poor were exploited by the rich. Paracelsus was not among their number. When he ventured into social ethics he emerged as a conventional Christian moralist, urging his readers, no less earnestly than Jesus, to set their hearts on 'things above'. Masters should be considerate in their demands and servants should not resort to force to gain their just desserts. While he leaned towards Anabaptist criticism of social evils, he did not support the use of violence to change the world nor the establishment of communes to withdraw from the world.

The one cognomen we *can* apply to this bundle of contradictions and the one that he readily acknowledged is 'magus'. This word, derived from the ancient Persian for an enchanter, an interpreter of dreams, a seeker of hidden (occult) truth, takes us back to the semi-mythical world of Hermes Trismegistus. Paracelsus certainly saw himself as belonging to a long tradition of 'adepts' who were not restricted by the assertions of Greek philosophy and medieval scholasticism. Magi sought, through the study of nature, to know the mind of God.

The late-fifteenth-century humanist scholar, Marsilio Ficino, defined the magus as 'a contemplator of heavenly and divine science, a studious observer and expositor of divine things'.* 'Contemplator' and 'observer' – that sums up the Paracelsian approach. This controversial figure stands somewhere between the Christian mystic and the empirical scientist. On the one hand he rejected anything that could not be demonstrated by experimentation. On the other he sought mystical enlightenment through studying the works of God

* Cf. C. Webster, *From Paracelsus to Newton*, CUP, Cambridge, 1982, p. 58

in nature. He was dismissive of black magic, the employment of arcane knowledge for evil ends (love potions, divining the whereabouts of buried treasure and the like) but was a firm believer in white magic (including transmutation) and the employment of arcane knowledge for the welfare of mankind. It is this dichotomy that lies at the root of his reputation. His writings, for the most part unpublished in his lifetime, were reverently printed by followers intent on stressing the contributions to science made by a man they revered as a genius. But to post-Enlightenment scientists he remains a charlatan who may have stumbled upon one or two useful discoveries but who was essentially an eccentric quack.

When we encounter such a unique thinker whose dogmatically asserted certainties we cannot understand, we may revere him as a profound philosopher who has penetrated truth at a level so deep that we can only wonder at his brilliance, or we may dismiss him as, at best, a muddle-headed crank or, at worst, a fraud. Whatever stand we take, we cannot escape from the fact that Paracelsus has become a legend and one that merits its place in the history of the pursuit of *scientia.*

CHAPTER SIX

Seeing may be believing

'I would rather hear that you were dead and buried, provided you had died in the Lord.'* The lady who expressed this opinion, having received unwelcome news about her elder son, Anthony, was Anne Bacon. What crime had the young man committed? He was sharing a room with a 'papist'.

Anne, the indomitable matriarch, was the widow of a leading English councillor and a prominent member of the court of Queen Elizabeth I. She was also an outspoken Puritan of very firm convictions. It goes without saying that she brought up her two sons 'in the fear of the Lord'. It probably also goes without saying that her sons were eager to kick over the traces as soon as they had escaped Anne's close maternal supervision. They completed their education at London's Inns of Court where a joke they told remained in circulation long after. They suggested that, after Mama's death, her ghost would return, wandering the corridors of Gray's Inn, wringing her hands and bewailing the loose morals of the students. Anne, sadly, lived to see her worst fears realised about Anthony. He was condemned as a traitor for taking part in the Earl of Essex's rebellion against the queen, though he died before sentence could be carried out.

It was left to Anne's younger son, Francis, to redeem the family honour. This he did – in spades. Winning the favour of James I, he rose to become Lord Chancellor and Viscount St Alban. However, in 1621 he was involved in a corruption scandal and forced from public office. The country's loss was scholarship's gain, for Bacon spent his

* J. Spedding, *The Life and the Letters of Francis Bacon*, Longman, Green, Longman and Roberts, London, 1861, pp. 110–111

remaining five years concentrating his considerable intellectual gifts in books which proposed a new approach to natural philosophy.

The strict Calvinism in which Bacon had been brought up had a rock-like, immovable foundation. 'No one,' the reformer wrote in his *Institution of the Christian Religion*, 'can have even the least taste of sound doctrine and know that it is of God, unless he has been to this school, to be taught by the Holy Scripture.'* The Bible contained everything necessary for salvation. It was to be understood literally, without the barnacle-like accretions that had been attached to it by generations of schoolmen. It was not to be obscured by analogies or allegorical readings. Certainly it was not to be used as a quarry of proof texts to support Church doctrines. Herein lies the secret of Bacon's approach to natural philosophy. God's book of nature should be read in the same way as God's book of Holy Writ. In his essay on 'Superstition', he paralleled the errors of Catholic theologians and philosophers wedded to Ptolomaic geocentricity:

> The schoolmen were like astronomers which did feign eccentrics [non-concentric orbits] and epicycles [the orbits of the near planets] and such engines of orbs to save the phenomena, though they knew there were no such things . . . in like manner the schoolmen had framed a number of subtle and intricate axioms and theorems to save the practice of the church.†

The Protestant reformers had democratised Christianity by making vernacular Scripture available to all and by insisting that it be understood simply. This 'decluttering' of biblical interpretation, as we have seen, went hand-in-hand with iconoclasm, the removal from churches

* J. Calvin, *Institution de la Religion Chrétienne*, ed. J. D. Benoit, J. Vrin, Paris, 1957, I, part 6, p. 2

† F. Bacon, *Essays*, OUP, Oxford, 1921, p. 47

of all 'objects of superstition'. The point of all this for the life of believers was not to gratify their intellectual curiosity, but to enable them to worship God 'in Spirit and in truth' (John 4:24). In other words, it was *practical religion*. No group took the 'stripping of the altars' more seriously than the Puritans. Bacon was brought up in the Puritan milieu and it provided him with the model for his philosophical reflections. To know and worship God more effectively it was necessary to 'read' Creation without preconceived ideas *and* to know that everything had been made by God for *use*, rather than speculation: the more man discovered 'the true nature of things', God would 'have the more glory' and man 'more fruit in the use of them'.* What followed from this was new initiatives in natural theology and, particularly, the development of the inductive method.

Aristotle's method of discovering truth, as set out in his *Organon* (*Instrument*), was to formulate hypotheses and then test them by observation. Bacon insisted that the philosopher must *begin* with observation and, by experimental testing, discover the truths to which it leads. Bacon lived in an intellectual world that was expanding at an unprecedented rate. Previously unknown lands were being explored and settled by European migrants. Harvey and the anatomists (see below) were probing the workings of the human body. Galileo was leading the way to more precise observation of the heavens (see p. 130). The old wineskins were unable to cope with all this new wine. The time had come for a fresh methodology.

Bacon set it out in his *Novum Organum* (*New Instrument*) of 1620. He was not an experimental scientist. William Harvey said of him that he wrote philosophy like a lord chancellor. There is, certainly, more than a breath of the law courts about his writings. A jury's responsibility was to lay aside any preconceptions about guilt or innocence, consider the facts presented to them and decide their

* F. Bacon, *The Advancement of Learning*, OUP, Oxford, 1974, p. 230

verdict on those facts alone. None knew better than Bacon that this was not always easy. Jurors could be swayed by honey-tongued counsel. The evidence presented to them might be incomplete. They might be confronted by witnesses who, through dishonesty, prejudice or sincere misconception, would present a distorted account of the facts. They might change their minds several times during the course of a trial. They might disagree among themselves.

Similar hazards made the natural philosopher's task complicated but Bacon insisted that, though the inductive method was not perfect, it was still the best available. Individual truth-seekers would sometimes make mistakes, just as the law courts sometimes freed the guilty or condemned the innocent. But the law was continually being refined; individual cases built up a corpus of precedent that gradually lessened the possibility of error. The same was true of science. The observations and analyses of scientists over the decades would tend towards an ever-improving understand of the awesome works of the Creator.

In his search for telling analogy, however, Bacon preferred the imagery of his Puritan upbringing. Calvin had demanded the purification of Christian worship and the removal from churches of all trace of 'idolatry'. In the same way, the *Novum Organum* insisted, the philosopher's mind must be cleansed from various kinds of idolatry. He identified four types of idols that tended to clutter the scholar's thinking space:

'Idols of the Tribe': human nature has inherited a distorted view of the world.
'Idols of the Cave': every individual is preconditioned by upbringing and education to see the world differently.
'Idols of the Market Place': daily interaction with other people colours our understanding. Even the words we use are loaded.

'Idols of the Theatre': earlier philosophers, using false or inadequate methodology, were like actors giving impressive performances which were, nevertheless, false.

We may see Bacon as the first proclaimer of scientific 'progress'. He believed that by corporate activity, aided by the best new technical inventions, thinking men would draw ever closer to the ultimate truth displayed by the Creator in all his works. Yet he certainly did not regard such progress as inevitable. Natural philosophers would always have to contend with the idols which distorted discovery and the theories based on such discovery.

A close contemporary, Johannes Kepler (1571–1630), probably never read Bacon's works but shared his motivation. As a Lutheran student studying at Tübingen he was planning to become a minister but his love of astronomy and mathematics set him on a new course. A 'vision' of the geometrical beauty of the universe in 1595 was, to him, almost like a conversion experience and he vowed to continue with his primary objective – to declare God's majesty and wisdom – but to do it by applying reason to his reading of the book of nature. As he stated later in his *Epitome of Copernican Astronomy* (1617–1620): 'It is a right, yes a duty to search in cautious manner for the numbers, sizes and weights . . . of everything [God] has created.'*

On leaving university Kepler took a post in Graz, teaching mathematics and astronomy. According to his students he was an infuriating lecturer. His mind was so full of ideas that he could not deliver a clear course of instruction. He frequently went off at a tangent, pursuing some new speculation. Such complaints are easy to believe when we come to consider his scholarly output. The collection of his works, currently being compiled, runs to twenty-two volumes, including subjects as diverse as planetary motion,

* M. Caspar, *Kepler*, Dover Publications, New York, 1993, p. 381

astrology and optics. He was also distracted by personal problems and the mounting pressure on Lutherans being imposed by the Habsburg government (see p. 123). It was his refusal to convert to Catholicism that eventually obliged him to leave.

He was drawn, with almost magnetic force, to Benátky Castle, near Prague, where Tycho Brahe was working on the finer details of his geo-heliocentric theory. He became the 'Great Dane's' assistant, with access to the mountain of accurate observations and measurements of the movements of heavenly bodies accumulated over the years. This proved to be Kepler's launch pad, for when Tycho died suddenly in 1601, Rudolph II appointed the newcomer to the post of Imperial Mathematician.

This privileged and, indeed, powerful position enabled Kepler to pursue his studies and publish his findings with the backing of his patron. His prime value to the emperor was as an astrologer. Europe's rulers still believed that the timing of important state events was dependent on the guidance provided by the movements of the planets. This was why Rudolph had backed Tycho's meticulous measurements and why he and his successors put their faith in his assistant. This backing gave Kepler the boldness to present to the world conclusions which he might otherwise have struggled to get accepted.

The logical process which would revolutionise astronomy began with his study of the orbit of Mars. He discovered that, for all Tycho's meticulous measurements, there was a discrepancy between where the planet was calculated to be and where it actually was. The flaw, as Kepler came to realise, was that his inherited data was based on the assumption held since classical times that the planets move in uniform, circular orbits. This was a prime example of Bacon's Idols of the Theatre. Once Kepler shifted the basis of his reasoning from unproven theory to observable fact, he concluded that Mars – and also the other planets (including Earth) – moved elliptically round the sun, which exercised some sort of 'pull' on their movements, giving them an ovaloid rather than a circular shape. Kepler had already become a Copernican. This discovery confirmed the heliocentric theory.

The solution to one problem brought other problems in its wake. What force held the planets captive to the Sun? Why did the speed of a planet's orbit change in relation to its distance from the Sun? Detailed mathematical calculation based on observation led to what came to be called Kepler's Laws of Planetary Motion:

1. Planetary orbits are elliptical, with the Sun as one of two focal points of their ellipse.
2. Although the planet moves faster when nearer the Sun, the radius of the ellipse in its sweep always marks out an equal area of space in any equal periods of time.
3. The square of the orbital period (the time taken to complete one circuit) is directly proportional to the cube of the semi major axis (the longest radius).

Kepler established two facts fundamental to all astronomical science: the beautiful mathematical precision governing planetary motion and the connection between mathematics and physics. Both pointed him towards one inescapable conclusion: inductive research supported the Christian revelation of a perfect mind behind the Creation and sustaining of the universe. As he has often been quoted as observing, 'We can only think God's thoughts after him.'

'Harmony' was, for Kepler, the key to the understanding of 'life, the universe and everything'. In 1611, before the invention of the microscope, he wrote a paper about snowflakes. Transferring his attention from the vastness of space to the minutiae of frozen water droplets, he described the beautiful and invariable geometrical precision of their 60-degree structures and sub-structures. He went on to speculate: what further structures might lie beyond what the human eye could see? If there were infinitesimal particles it was reasonable to suppose that they, too, must obey the same mathematical laws. Herein lay the first footprints on the dusty road leading to atomic theory.

Eight years later, Kepler expanded his observations about the way mathematics illuminates the purposes of God. *Harmonices Mundi* was a treatise in five books about the geometrical precision of the divine mind as revealed in astral movements, music, architecture and pure mathematics. Man, he asserted, responds psychologically and aesthetically to the primordial principles of harmony because he is made in the image of God. Medieval musicians, theologians and stone masons had stated or, at least, felt the imperative of divine harmony. Kepler was now bringing inductive reasoning to bear on the same phenomenon.

Science populariser Carl Sagan said that Kepler was 'the first astrophysicist and the last scientific astrologer'. His continued enjoyment of imperial patronage and a substantial part of his income came as a result of the production of horoscopes for the emperor and other wealthy clients. But he was uneasy about this aspect of his work. Astrology did not sit easily with either his understanding of astral phenomena or his Lutheran faith. He held firm to the traditional belief that planetary motions and alignments had direct influence on terrestrial events and, therefore, on the lives of individuals but he was sceptical about judicial astrological predictions that claimed to offer specific advice on personal or political problems. He believed that the harmony he discerned in the heavens must relate to the formation of character and resultant human behaviour but to regard the influence of the planets as deterministic would be to limit both divine sovereignty and human free will. He would certainly have agreed with Shakespeare's Cassius in *Julius Caesar*, 'The fault, dear Brutus, is not in our stars, but in ourselves, that we are underlings.' Yet it was the influence of the stars that most of Kepler's customers wanted to hear. Astrology paid the bills and he could not afford to give full rein to his scepticism – and, in fact, his predictions owed as much, if not more, to his acute observations of political, religious and military events as they did to his casting of horoscopes.

It was hard for any scholar in the public eye to maintain integrity at a time when political and religious forces were pulling in several directions. Kepler was watched closely by papal agents at the court of the Catholic emperor and he fell under suspicion with the Lutheran top brass because of his inclination towards certain Calvinist tenets (particularly with regard to Eucharistic theology). Kepler experienced the fundamental difficulty of the Christian scholar. His mind had to be free to go wherever reason led but his liberty did not extend to challenging divine revelation. As far as natural philosophy was concerned, reason was king and free to question traditional authority but in matters of faith the authority of the Bible was sovereign. In any case, Holy Writ was a spiritual guide, not a scientific textbook. Yet it was easier to state these principles than to actually live by them.

As with other influential thinkers, we have to understand Kepler in the context of the events that dominated his world and his life. He is not a rigid milestone on some imaginary road from religion to reason; superstition to science. His passion for harmony and unity was, in large measure, his response to contemporary disharmonies and disunity. Kepler's life coincided with one of the most wretched periods in European history. Preachers of all religious stamps prophesied the imminent end of the world and, although such apocalyptic warnings were not unprecedented, they did seem to have particular relevance as the Christian West lurched into the new century.

The 1590s afflicted the Continent with prolonged meteorological disaster. Crops failed. Famine was widespread. At the same time the conflict between Reformation and Counter-Reformation really got into its stride. The Dutch bid for independence from Catholic Spain that produced the series of military encounters known as the Eighty Years' War from 1568 had produced an exhausted stalemate that led to a long truce, but hostilities resumed in 1621. They morphed into the more wide-ranging Thirty Years' War (1618–1648) which devastated northern Europe and wiped out 35 per cent of the populations of the combatant nations (besides which the death toll of the two European wars of the twentieth century pales into insignificance as a proportion of total citizens). As Professor Diarmaid MacCulloch points out, those caught up in these disasters might well have believed that the first two Horsemen of the Apocalypse (Famine and War) were already imprinting their hoofmarks on the devastated land.*

But religious and political dislocation struck Kepler closer to home. Graz, Austria, where he lived during the last years of the old century and where he gained his first paid employment, was one of the nubs of confessional conflict. It was one of the few cities in the country where Lutheranism had taken a firm hold. It was also the capital of the archdukes of Inner Austria who were leading figures in the Counter-Reformation. But the Archduke Ferdinand, in close concert with the papacy, was determined to abolish the religious freedom acquired by the Protestants (due largely to their commercial leverage) and he did so by a steady process of attrition. Jesuits were brought in to establish a rival school to the one run by the Lutherans. In 1598 they were granted an educational monopoly and it was the closure of his own school that forced Kepler to leave.

It was not only Catholic intransigence that complicated Kepler's life. In 1613 the emperor sought his advice on the adoption of the proposed Gregorian calendar. Over the centuries the calculation of important dates, such as Easter, had been based on the Julian calendar which, as Jesuit astronomers explained, was, by then, ten days

* D. MacCulloch, *Reformation: Europe's House Divided 1490–1700*, Penguin, London, 2003, p. 554

out of sync with accurate measurement. Pope Gregory XIII had decreed that there should be an immediate adjustment by removing ten days from the prevailing calendar. Kepler gave his wholehearted support to the change in the interests of scientific accuracy. He now found himself at odds with his Lutheran friends who resisted the 'Catholic plot'.

For decades or, in some places, centuries, confusion reigned. Travellers crossing state boundaries had to change their diaries. Those arranging appointments in distant towns had to know what system operated in the place they were going to. Worse still, in war-dislocated seventeenth-century Europe, territorial gains and losses automatically involved calendar changes. To a man of Kepler's precise turn of mind this must have been intolerable and yet another reason for seeking harmony and unity.

Repeated disruptions of his own circumstances affected Kepler's thinking and we might well marvel that he was able to produce such a prodigious volume of work. Increasingly uncomfortable in Prague, he applied unsuccessfully for posts elsewhere and finally moved to Linz in 1612. At the same time his first marriage, which was not a happy one, came to an end with the death of his wife. When, in 1613, he made a more agreeable match it was to see his first three children die in infancy.

Between 1617 and 1620 he was involved in his mother's problems. She had been labelled a witch by a vindictive neighbour and Kepler had to fight her cause in the courts. Archduke Ferdinand, who had begun the persecution of Lutherans in Inner Austria, became Emperor Ferdinand II in 1619, determined to eradicate every vestige of Protestantism from his wide dominions. One result was that aspects of Kepler's work tended to be viewed with suspicion by Catholic scholars. In 1625 his library was sealed by Counter-Reformation agents. Yet, at the same time, his refusal to swallow official Lutheran doctrines that he could not, in reason, accept, meant that he was excluded from holy communion in his own church. As if such major conflict was not enough, in 1626, Linz came under siege from rebellious peasants and Kepler was obliged to move his family to Ulm.

However much he tried to concentrate on his teaching, research and writing, he could not escape political involvement. His astrological calculations continued to occupy his time, the more so as the fluctuating fortunes of war brought demands for predictions from politicians and military leaders.

In 1528 he was appointed an adviser to the court of Count Wallenstein, the imperial general in charge of Catholic forces in the seemingly endless war. He was now caught up in ceaseless travel between Prague, Linz, Ulm, Sagan (Wallenstein's occasional headquarters) and Regensburg. This was not just tiresome; it was dangerous: in 1626,

> hunger and pestilence had accounted for twenty-eight thousand. Disease could not be checked with the armies passing; typhus, scurvy, smallpox, syphilis, marched under the banners and bred in the countryside. Diseased horse and cattle trailed along among the baggage wagons, spreading contagion in the farms through which they passed.*

What sickness Johannes Kepler succumbed to is not known but, with all this disease and mortality he may have been fortunate to survive until November 1630. Even after his death the dislocated times had not finished with him. A victorious army rampaged through the graveyard where he lay buried, obliterating for all time his memorial. Much of his work was similarly lost to view. Not until several decades had passed did the scientific world begin to appreciate the achievements of

* C. V. Wedgwood, *The Thirty Years War*, Jonathan Cape, London, 1947, p. 217

this unassuming scholar who patiently went about his work, not boasting of his discoveries or arrogantly challenging those who disagreed with him, but singing always a song of harmony and concord to a world that chose not to listen.

We ought to be thankfully aware of the not inconsiderable number of scholars who beavered away at extending human knowledge but whose importance was only realised years, or even centuries, later. One such was Thomas Harriot (1560–1621), in whom we have an early example of a species just beginning to evolve – the amateur scientist. Men who either possessed the means to indulge their scholarly hobby, or who enjoyed the support of wealthy patrons, made their own calculations and discoveries and discussed them with others who shared their interests. Scholars had always exchanged ideas and engaged in debate through correspondence. Thanks to the intellectual *lingua franca* of Latin, an international network had emerged in the Renaissance, linking English, French, German, Italian, Spanish and Dutch-speaking academics who traded their thoughts on a range of philosophical and theological issues.

Most of the participants were university-based but the times were changing. Non-academic, self-trained enthusiasts were now joining the 'experts'. First in Italy and gradually in lands beyond the Alps, princes, bishops, wealthy merchants and even emperors gathered around them men whose reputations as advanced thinkers added lustre to their courts. But now a new species of learned society was emerging that was not housed in universities or dependent on rich patrons.

In Rome the Accademia dei Lincei advertised for members who were 'eager for real knowledge and will give themselves to the study of nature, especially mathematics'. By 1620 similar groups were meeting in Florence, Venice, Madrid, Paris and other cities. From such beginnings England's Royal Society and other similar bodies developed in the later seventeenth century.

It was around 1580 that Harriot emerged from Oxford with a reputation as a bright, multi-talented scholar intent on continuing his studies. He was fortunate to attract the patronage of Elizabeth I's court favourite, Sir Walter Raleigh. Raleigh was, at this time, much engaged in pioneering colonising ventures in North America and he found Harriot useful in several ways. As well as mathematics and astronomy, the young scholar was something of a linguist. When an exploratory expedition returned from what Raleigh named Virginia, the leaders brought with them two Algonquian Amerindians. It was Harriot who spent time with them, learned their language and devised a phonetic alphabet to render it printable.

Between 1585 and 1586 Harriot was part of an ill-fated attempt to establish a colony on Roanoke Island, in what is now North Carolina in the USA. There he continued his studies of this strange 'new' land and people. He explored and described the flora and fauna of the region, always with an eye to discovering commodities that could be exploited for profit. The American interior seemed to be very extensive and, just as the Spanish *conquistadores* had found veins of precious metal in their part of the New World, so it seemed a virtual certainty that intrepid exploration by Englishmen would reveal marketable minerals or vegetable resources. They would also, Harriot observed – like many other imperialists, then and later – be performing a service for the natives by bringing them to 'civility and the embracing of true religion'.

Harriot was a practical, rather than a theoretic scientist. The patrons who paid – and paid handsomely – for his advice had political and economic ends in view. Raleigh, for example, had received from the queen grants of lands in newly acquired territory from which he intended to profit by selling individual holdings to settlers. Like John Dee, Harriot was interested in the expansion of England's maritime enterprise and, as a true patriot, in challenging Spanish and Portuguese imperialism. He instructed Raleigh's mariners in the science of astronomical calculations. His quest for better navigational instruments led to the development of the four-metre-long backstaff which, though unwieldy, yielded the most accurate astronomical readings available at that time. On his return, Harriot published *A Brief and True Report of the New Found Land of Virginia*, a promotional treatise to encourage potential settlers and investors which, among other enticements, advocated the pleasure of smoking tobacco.

One of the hazards of the patronage system was that a protégé might rise with his 'good lord' when the latter's star was in the ascendant but could decline into obscurity – or worse – if the patron's star waned. And, towards the end of Elizabeth's reign, Raleigh fell from favour and the new king, James I, so far 'took agin' the courtier that he had him incarcerated in the Tower of London.

Harriot escaped being adversely affected because he had been taken up by another forward-thinking patron, Henry Percy, Earl of Northumberland. The earl set him up in a house on his estate near London with a generous pension and carte blanche to pursue his studies. Sadly, it was not long before his new protector also encountered royal displeasure. Percy was accused of marginal involvement in the Gunpowder Plot of 1605. He, too, was consigned to the Tower, at His Majesty's pleasure. Harriot also fell under suspicion, was briefly imprisoned, interrogated and had his library searched for incriminating evidence. He avoided the taint of 'treason' but his reputation suffered from an equally damning crime – 'atheism'.

Today the charge against him would be better rendered as 'scepticism'. Harriot was identified with a group of freethinkers who gathered around Raleigh and Percy to discuss the latest developments in science. These disgraced ex-courtiers were not shut away from the world in cramped cells living on bread and water. Far from it; they enjoyed five-star luxury in spacious quarters, where they could receive visitors and entertain in some style. All they had lost was their liberty.

Percy (known to some as the 'Wizard Earl') lived in almost regal splendour, dining off silver and gold plate, his rooms equipped with fine furniture, hangings, carpets and paintings and his table furnished with food prepared to his own specification. Both men had still houses built where they could conduct alchemical experiments. A third member of their imprisoned fellowship, Lord Cobham, had a special turret built to accommodate his library of a thousand books.

There can scarcely ever have been a stranger setting for a cultured salon, yet that was exactly what these royal prisoners gathered and established and Thomas Harriot was its leading light. What is not surprising is that this coterie of prisoner-scholars was viewed with suspicion. The fact that the senior members of this intellectual coterie were fallen stars must have had much to do with the dubious reputation the group acquired. It was referred to in respectable quarters as the 'School of Night', whose members dabbled in forbidden knowledge. The stigma long survived. Towards the end of the seventeenth century John Aubrey, in his gossipy *Brief Lives*, wrote of Harriot that he

> made a philosophical theology, wherein he cast off the Old Testament and, consequently, the New one would have no foundation. He was a Deist. His doctrine he taught to Sir Walter Raleigh, Henry Earl of Northumberland, and some others.*

We are back in the world of *Dr Faustus* and forbidden knowledge, and it is no surprise to learn that Christopher Marlowe was well acquainted with Harriot. In regarding the Tower 'academy' with suspicion, the fashionable world took its cue from King James, who

* J. Aubrey, *Brief Lives*, ed. A. Clark, (1898), Penguin Books, London, 1972, p. 287

was almost paranoid in his hatred of witches and everything that smacked of 'magic'.

But what are the facts? Harriot's contributions to science are difficult to assess because he published very little. This might have been because he was diffident by nature or because he was wary of presenting novel ideas to the public for fear of being misunderstood (a concern we have already encountered in other scholars). Though a fascinated devotee of several branches of *scientia*, Harriot was basically a mathematician/astronomer. He pioneered developments such as the use of algebra, an accomplishment only appreciated a decade after his death when Walter Warner (another member of the School of Night) published some of his papers under the title *Artis analyticae praxis* (1631). He was a convinced Copernican and actually anticipated Kepler (with whom he was in correspondence) in asserting the elliptical orbits of planets.

It was, doubtless, his connection with political men of the world that helped Harriot to keep abreast of what was being done and thought by scholars in other countries. Thus, when he heard, in 1608, that a spectacle maker in Holland had invented something called a 'Dutch trunk', he was quick to acquire a specimen. Hans Lippershey had devised an arrangement of lenses within a tube that enlarged the appearance of distant objects by a factor of three. The inventor obtained a patent for this proto-telescope as a useful tool for the Dutch military in their long-running war with Spain but it was Harriot who grasped its astronomical potential.

He used it first to explore the moon and made the first drawing of the lunar surface. Subsequently, he produced and distributed to friends improved telescopes (eventually offering x50 magnification). As well as accurately delineating the moon's surface, he observed the satellites of Jupiter and sunspots. In these years Galileo and other astronomers were also exploiting the exciting possibilities of the telescope (see below) and there is still debate over exactly who was the first to make various discoveries. But Harriot seems not to have been interested in academic competition. He freely exchanged information with several correspondents. This and the fact that he did not bother to publish most of his findings suggests a lack of ambition.

On the other hand, his reticence about making a name for himself may have had something to do with a desire to avoid the reputation of being part of an atheistic cell. He shunned scholarly controversy. For example, his correspondence with Kepler suggests that he rejected astrology *in toto* but we are left to speculate about that; no clear statement about his attitude to the influence of heavenly bodies has survived. When Harriot died at the age of around sixty-one he was accorded full Christian burial in the London church of St Christopher-le-Stocks. All trace of his memorial disappeared in the 1666 Great Fire of London, although the Latin inscription survived in John Stowe's *Survey of London*. The epitaph extolled Harriot as one 'who treasured all knowledge and excelled in mathematics, philosophy and theology'. He was a 'most studious enquirer into truth and a most pious worshipper of the Trinity'. The wording suggests a determination to distance Harriot from any suggestion of atheism or deism. We must charitably assume that the deceased intended this assertion to be taken at face value.

By no stretch of the imagination could Harriot's contemporary, Galileo Galilei (1564–1642), be called a modest scholar or content to pursue his studies with little thought for personal reputation. In his last published work, *Discourses and Mathematical Demonstrations Relating to Two New Sciences* (1638), he introduced his theme in these words:

> My purpose is to set forth a very new science dealing with a very ancient subject. There is in nature, perhaps, nothing older than motion, concerning which the books written by

> philosophers are neither few nor small: nevertheless, I have discovered by experiment some properties of it which are worth knowing and which have not hitherto been either observed or demonstrated.*

His treatise was written when the author was technically under house arrest and was published in Holland because Italian printers were too afraid to touch it for fear of the Inquisition. Yet within months the book was a bestseller in Rome and Galileo seems not to have attracted unwelcome attention from the Vatican. If this appears paradoxical it is because we have become accustomed to regarding Galileo as a daring thinker of new thoughts who came to grief because he lived and worked at a time when the Catholic Church was determined to suppress heretical 'novelties'. Such a black-and-white interpretation needs challenging.

Galileo was born in Pisa, the eldest child of a talented musical family. His father, Vincenzo Galilei, was a fine lutenist, an innovative composer and a writer of musical theory. His younger son, Michelagnolo, would become an equally acclaimed musician. Galileo's family was cultured and, though not poor, had probably seen better days. He was certainly impressed by his father with the need to get on in the world so as to be able to contribute to the domestic finances. It was probably for economic reasons that Vincenzo moved to the great cultural hub of Florence while his eldest was still a child. He sent the boy to a monastery school where he soon displayed a religious bent. He joined the novitiate. But Vincenzo soon put a stop to that. It was necessary for Galileo to contribute to the family economy. The boy had a good head on his shoulders and needed to be put to a more lucrative career. And few careers were more lucrative than medicine.

* Galileo Galilei, *A Discourse Concerning Two New Sciences*, Dover Books, New York, 1954, p. 153

The dutiful teenager enrolled at Pisa University and began the dreary study of Galen's standard texts. But he soon discovered something more to his taste – natural philosophy and, especially, mathematics. According to legend, Galileo's new passion began with a chandelier. He observed the massive lamp in the cathedral swinging in the breeze. How many thousands – millions – of visitors had seen this before and never given it a second thought? For some reason Galileo decided to time the motion (using his own pulse as a clock). What he discovered was that, no matter how long or short the arc, the natural pendulum took the same time to complete it. The story may be apocryphal but it points us to two fundamentals of Galileo's mature thinking: the process began with observation and it could be understood with the aid of mathematics.

It is not too fanciful to see the origin of Galileo's truculent individualism in his early attempts to establish his own identity. His determination to challenge his father's career choice and his distaste for the slavish following of ancient wisdom in the medical school were early pointers towards the rejection of ideas he could not prove for himself. In a work of 1632 he lampooned the tame followers of Aristotle who trawled the philosopher's works:

> That from divers passages they may quickly collect and throw together a great number of solutions to any proposed problem . . . as if this great book of the universe had been written to be read by nobody but Aristotle and his eyes had been destined to see for all posterity.*

But that was many years off when the young student first became enthralled by 'the book of nature'. 'The glory and greatness of

* Galileo, *Dialogue Concerning the Two Chief World Systems*, trs. S. Drake, University of California Press, Berkeley, 1953, p. 18

Almighty God,' he declared, 'are marvellously discerned in all his works and diversely read in the open book of heaven.'

Galileo was a polymath, much in the vein of da Vinci. He was an accomplished musician and well versed in the plastic arts. His interest in applied mathematics led him to invent various precision instruments of practical use to artisans. He was well known in the Florentine artistic coterie and, at the age of twenty-four, became a teacher of perspective and chiaroscuro in the Academy of Art and Design in the city that still led the way in late-Renaissance culture. The following year he moved back to Pisa to take up the chair of mathematics.

This CV may seem to the modern reader to indicate a talented and successful scholar rising rapidly through the ranks of academe, but Galileo's stipends were far from generous, he did not enjoy teaching and he was not helped by his poor management of domestic accounts. He had a reputation for sponging off friends and his frequent bewailing that his genius was not recognised certainly tried their patience.

In an age when patronage was the most secure ladder to fame and fortune he commended himself to the Grand Duke Ferdinand I of Tuscany but without much concrete success. On the face of it this is strange. Ferdinand was the most enlightened member of the Medici family since the great Lorenzo the Magnificent. As well as ruling wisely, encouraging all aspects of art, including the new musical form called 'opera', bringing to the Uffizi his own extensive collection of classic artefacts, building the finest navy in the Mediterranean and boosting Florentine commerce, he developed Livorno into a flourishing, deep-water port, connected to Pisa by a navigable channel through newly drained marshland. He was deeply interested in mechanics and engineering and he had a high regard for Galileo, calling him 'the greatest mathematician in Christendom'. Yet he did not help the struggling scientist with lucrative employment. Could it have been Galileo's reputation for arrogance that, at this early stage, hampered his career? Ferdinand did employ Galileo's services for a brief spell as tutor to his son and heir, Cosimo, and this would bear fruit at a later date but for now Galileo had to seek pastures new.

In 1592, he took up a post as lecturer in geometry and astronomy at the University of Padua. But he certainly was not forgotten in Tuscany. In 1609 he received an urgent message from Grand Duchess Christina. Her husband was grievously ill and she wanted the great 'astrologer' to cast his horoscope. Galileo eagerly obliged, assuring Ferdinand's wife that he would soon recover and live to a ripe old age. The duke was dead within the month.

The Paduan years were frustrating for Galileo, although he did not lack widespread acclaim as an inventor and expositor of natural philosophy, nor was his environment hostile to exciting new ideas. The university was under the aegis of Venice, a state no less vigorous than Tuscany in its encouragement of scholarly innovation. He had many friends and correspondents in the wider intellectual world. Yet he felt undervalued. Aristotelian science – in Galileo's mind outmoded and inadequate – still held sway in the upper echelons of the academic hierarchy. His domestic life brought him little comfort. He remained technically celibate, in order not to hamper his chances of academic preferment but, like many in his position, he lived with a mistress, Marina Gamba, and by her had three children. He also had to support his younger brother, Michelagnolo, and his two unmarried sisters after their father's death in 1591. His financial problems were a constant worry. In short, Galileo Galilei was not the first or last genius to feel alienated from a world that did not understand him or accord him that status he considered to be his due.

He had certainly done enough to attract fame, if not fortune. He had devised a hydrostatic balance as an aid to bankers and jewellers in weighing and determining the purity of precious metals. He invented a 'compass', an instrument of value to surveyors and military engineers enabling the easy calculation of geometrical angles. He produced an implement for indicating temperature differences, a forerunner of the thermometer. All his studies in practical mathematics were contributions to the solution of problems engaging many scientific minds around the turn of the seventeenth century and Galileo was in correspondence with thinkers in several countries. It was, therefore, inevitable that he would be among the first to hear about the 'Dutch trunk'.

News of Lippershey's invention reached Venice early in 1609. Within weeks Galileo had pounced upon the opportunity to turn this new phenomenon to his advantage. He grasped the principle involved and made his own version, an instrument providing eight-fold magnification. This frenzied activity was necessary because someone else had turned up in Venice hoping to sell a similar invention to the government. The age-old dictum, 'What matters is not what you know but who you know,' now came to Galileo's aid. He had friends in the city with whose help he gained recognition by the authorities. He generously presented his 'telescope' to the senate and in return the government appointed him professor for life at Padua with an increased salary.

The Venetians were impressed with the military potential of the new invention but Galileo immediately began stargazing with it and he was unaware of Harriot's initiative when he began examining the moon. Though he sketched what he saw, he did not make a detailed map of the lunar surface. This may have been because he had, by then, seen the Englishman's precise drawings or because he had moved on to other and more exciting discoveries. He closely observed Jupiter and Venus and realised that they were sun-orbiting planets, like Earth. Jupiter, indeed, had four satellites of its own.

The year 1609 was a turning point in Galileo's fortunes in more ways than one. The death of Ferdinand I brought to the throne of

Tuscany his old pupil, as Cosimo II. Galileo was not slow to take the opportunity of turning his back on Padua. He hastened to the press an account of his recent discoveries (*The Starry Messenger*) and dedicated it to Cosimo. As further bait, he named Jupiter's moons the Medicean stars. Whether or not as a result of these tactics, Galileo struck gold. He was appointed official mathematician and philosopher to the court of Tuscany with a new villa built at Arcetti, just outside the city and close to Cosimo's own palace. Here he was often visited by the grand duke and his eminent guests, who were enthralled by what they were able to observe through the telescope. Galileo had *arrived*. He was a celebrity, famed throughout Europe and feted by the great and good. The time had come to disembarrass himself of anything that might damage his reputation. Marina was packed off to her relatives in Venice.

Over the next few years Galileo improved his instrument (eventually achieving x30-plus magnification) and directed it at different parts of the heavens. He discovered that the Milky Way is composed of an incalculable number of stars and planets. He observed sunspots on the Sun which demonstrated its rotation. He detected what we now know as the rings of Saturn, though he could not see them clearly enough to understand their true nature. Everywhere he looked in the night sky he discovered a universe vaster, stranger and more wonderful than anything human beings had ever witnessed or even imagined.

Then he came down to Earth and had to try to explain what he could see to bigoted dullards who could not be shaken out of their blind faith in ancient 'wisdom'. The longed-for evidence now existed to settle some of those issues about which philosophers had long argued – particularly, 'What goes round what?'

Galileo was by now a convinced Copernican. What the Polish scholar had *suggested* on the basis of mathematical calculation he could now *demonstrate*. Never a man to suffer fools gladly, Galileo was impatient with opposition. To him it was crystal clear that if the Aristotelians continued to reject heliocentrism they were simply clinging to old – bookish – error and refusing to face up to new – evidential – truth. Characteristically, he was unsparing in his contempt. In a letter to Kepler he mocked as 'deaf asps' those critics who had even refused to look through his telescope. We today share the writer's impatience with such closed-mindedness but we need to try to put ourselves in their position. What they were being offered in Galileo's strange tube was something absolutely revolutionary, something outside the scope of their intellectual discipline – something that smacked of *magic*. They looked at Galileo and, behind him, saw the hovering shade of Faustus. Galileo dismissed such concerns. Unfortunately, he failed to distinguish between philosophical and theological opposition. Since Greek reasoning had for centuries been seen as lending support to Christian truth, Galileo's contempt for book-bound Aristotelians could look like an attack on the doctrinal authority of the Catholic Church. At a time when the Counter-Reformation was at its height, it seemed to some ardent defenders of the faith that, though the bumptious Florentine celebrity was undoubtedly a genius, he was getting ideas above his station.

Sure of his position, in 1611, Galileo travelled to Rome to demonstrate his telescope and discuss the phenomena he had observed with Jesuit scholars, the Pope's specialists in natural philosophy. His reception was polite but mixed. He was invited to join the Academia dei Lincei and members who had made their own telescopic observations verified Galileo's discoveries. This did not, however, mean that they accepted his interpretation of those discoveries. Some clung to Tycho Brahe's geo-heliocentric explanation of planetary motion. Others refused to accept any theory that removed Earth from the centre of God's Creation. Then there were those who were jealous of the Florentine's success or who simply took a dislike to his manner.

'Beware of the Pigeon League'. The warning came from Galileo's closest friend in the city, the artist Lodovico Cardi (known as Cigoli) who was currently working on a remarkable fresco on the dome of Santa Maria Maggiore. It depicted the Madonna standing on the moon – the moon as revealed by Galileo's drawing, all pockmarked and irregular. The Pigeon League was his code for a malicious group ready to use any excuse to make trouble. There were many such in the bickering, backbiting, claustrophobic little world of the Eternal City.

Back in Florence the atmosphere was calmer but the issue of Copernicanism versus Aristotelianism and literal Biblicism was still in the air. In 1613, he was called on to advise the widowed mother of his patron, Cosimo II, who was concerned about the apparent contradiction between heliocentrism and the plain words of Holy Writ, particularly the claim in Joshua 10 that God made the Sun stand still for a day. By way of response Galileo wrote *A Letter to the Grand Duchess Christina.* His basic attitude was, as the Church historian, Cardinal Cesare Baronio, had once said to him, 'The Bible teaches us how to go to heaven, not how the heavens go.' Where the word of God described natural phenomena, it was legitimate, as Augustine had suggested, to read it allegorically. Not for a moment did Galileo abandon his belief in God's 'two books'. Scripture and Nature were complementary, serving different purposes. However, he did offer a scientific answer to the Joshua 10 problem: it might be that God had removed sunspots, thus giving the impression that the sun seemed to have stopped. It was not the most convincing 'get-out-of-jail' card, as his enemies now gleefully observed. They were not slow to pounce.

The opening salvo was fired from St Mark's Dominican convent in Florence. Towards the end of 1614 (before *A Letter to the Grand Duchess* had even been published in its final form) one of the brothers, Tommaso Caccini, preached a vitriolic sermon denouncing

Galileo and the whole tribe of 'disobedient', 'free-thinking' mathematicians. Caccini was a loud-mouthed extremist deliberately courting fame by his fiery sermons. Like many orators throughout history, he built his career on emotional outpourings that were a triumph of style over substance. Right-thinking men were appalled by Caccini's fanatical condemnation. One of the preacher's Dominican superiors wrote a letter of apology to Galileo. But the accuser thrived on notoriety. He drew the attention of the Inquisition to the errors of Galileo and his supporters who he said were interpreting the Bible in the light of their own distorted reason in flagrant opposition to the teaching of the Church.

Despite poor health and against the advice of friends, Galileo travelled to Rome to face his accusers and to elicit the help of those in the city he knew to be sympathetic towards him. He gave no quarter in his arguments supporting Copernican theory and his characteristic truculence cannot have helped his cause. But the decision of the Holy Office was a foregone conclusion. In February 1616, they reported to His Holiness that heliocentrism was contrary to reason and Scripture. Paul V, therefore, ordered Galileo not to teach or defend such doctrine. At a personal audience the Pope assured Galileo of his personal admiration and promised him that he had nothing to fear as long as he, the Pope, lived.

Pope Paul might have seemed friendly and tolerant but there was another side to his nature that could be harsh and even homicidal when it came to defending papal authority. This had been exposed a few years earlier in the shocking affair of Paolo Sarpi, a prominent Venetian monk, philosopher and scholar. In 1606 a dispute arose between the Pope and the Venetian government concerning their respective legal rights. Sarpi, an expert in canon law, presented learned arguments supporting the supremacy of civil courts in secular matters. Paul responded by excommunicating the Republic of Venice. The Venetians' reaction was to exile all Jesuits (the prime defenders of papal authority). In the midst of all this, Sarpi was engaged in a vigorous pamphlet war with his opponents.

An enraged Paul V was determined to silence this presumptuous monk. In September 1607 he paid three ruffians eight thousand crowns to assassinate Sarpi. The plot was discovered and the would-be murderers were imprisoned. However, another attempt a few weeks later was almost successful. Sarpi was left for dead with fifteen stab wounds by villains who escaped into papal territory where they were warmly welcomed. Fortunately, Sarpi recovered and lived to write further critiques of papal power – notably a monumental *History of the Council of Trent*, which he denounced as having made the name of the Catholic Church widely hated. Compared with such scandals the Galileo affair pales into insignificance.

Before returning home, Galileo obtained a certificate declaring that he had not been ordered to recant or do penance. He could show the world that he was still a good son of the Holy Church. At least some of the mud thrown at him had not stuck. In fact, the official position regarding his scientific opinion, as defined by a leading Vatican theologian, Cardinal Bellarmine, was that heliocentric theory was fine as a hypothesis, a mathematical device, that 'saved the appearance' of things – i.e. that it agreed with the observable evidence. It could be produced in academic debate as a means of gaining further understanding but it was not to be defended as scientific truth because it flew in the face of revealed truth. Copernicus' work was placed on the Index, but only as a treatise requiring some amendment. This was, in fact, far from being a foolish, dog-in-a-manger rejection; Galileo's findings on the relative movements of sun and Earth had not been *proved* beyond doubt. It was still possible for other astronomers to come up with their own explanations of what was visible through the telescope.

For the next few years Galileo was able to continue unmolested with his studies, making fresh observations (e.g. on the nature of comets) and engaging in robust controversy with his peers. His world changed considerably in those years. In 1618 central Europe disintegrated into three decades of devastating politico-religious war. Closer to home, Cosimo II died in 1620, at the age of thirty, to be succeeded by a minor. No longer did Galileo have the steady support of a sympathetic and pro-active patron. Death also removed Paul V. This brought to the papal throne someone who might have been even more valuable to the ageing scholar. The Florentine Cardinal Barberini who now became Pope Urban VIII was an old friend and a fellow member of the city's scientific academy. Yet this was the man who has gone down in history as the arch-enemy of science.

At the time Urban's persecution of Galileo was not regarded as one of the outstanding events of his reign. In Rome it was said of him, 'What the barbarians did not do, the Barberini did'. He was a lavish builder, leaving to posterity the Bernini colonnade around the Piazza of St Peter's, the baldacchino under the cathedral dome and various palaces. The building materials for these projects came from wholesale filching of what remained of classical Rome. That was not all Urban stole. He was the last pope to make large extensions to the Papal States by laying legal claims to various territories in northern Italy, sending papal troops to enforce those claims and installing members of his family as rulers. His was nepotism on a scale that had not been seen since the Borgias. One of the duchies towards which he extended his claws was Urbino, which had come by marriage to Cosimo II's sister. Understandably there was bad blood between the Medicis and the Barberinis. This leitmotif was to play steadily during the last great drama of Galileo's life.

It began well. Galileo went to Rome and presented to his old friend the latest fruit of his genius, a book called *The Assayer*. It was an explanation and defence of the scientific method, displaying how truth is approached by observation and observation is interpreted by mathematics. Mathematics is, indeed, the language of science. *The*

Assayer, dedicated to Urban, mercilessly attacked the Aristotelians and was hugely enjoyed by those who stood on Galileo's side of the fence. But for those who did not – well, this was another example of the author's skill at making enemies. Urban had the book read to him at meal times and appreciated his old friend's witty (and not so witty) barbs. He now encouraged Galileo to set out clearly in print exactly what it was that the Copernicans and the Aristotelians were arguing about. His only caveats were that the book should include Urban's own teaching and that it should not advocate heliocentrism. In 1632, Galileo accordingly published, with the Pope's blessing, his *Dialogue Concerning the Two Chief World Systems*. That was when everything went pear-shaped.

Whether the author did not understand Urban's conditions or whether his hubris fatally got the better of him has never been clear. What is beyond doubt is that Galileo's erstwhile friend became his enemy. The Pope felt that he has been disobeyed and betrayed. And he had every justification for being offended. The new book took the form of a debate between 'Salviati', a Copernican spokesman, and his Aristotelian counterpart, 'Simplicio', in the presence of 'Sagredo', a witty referee. Now, after years of frustration and of patiently enduring the taunts of his enemies, Galileo opened the sluice and let his feelings flood out.

Simplicio was presented as a buffoon whose feeble arguments were mercilessly exposed by the other disputants. Heliocentrism emerged as the only workable hypothesis. That alone would have been enough to get the author into trouble, for he had flouted the Pope's interdict. But what was worse was that the arguments put in the mouth of Simplicio were precisely those Urban held and had ordered Galileo to include.

By 1633 Urban had been head of the Catholic Church for ten years and had established himself as a corrupt, implacable dictator. He was widely unpopular (certainly in Florence) as a pope more concerned about glorifying his family than defending Catholic truth at a time when much of Europe was convulsed by conflict over that truth. Only four years later he had to stave off a rebellion by the College of Cardinals. All this forms part of the background to the incident for which history tends to remember him. It helps to explain why Urban was so sensitive to any challenge or perceived challenge to his honour and authority. What it does not explain is Galileo's insensitivity. He was simply a genius who seldom troubled to look outside the ivory tower of his own intellect.

The result of this clash of egos is well known and needs no detailed retelling. Galileo, now entering his seventies and much afflicted by poor health, was summoned to Rome to answer to the Inquisition for his contumacy. No scientific evidence was put forward and no scientific truth specifically challenged. The prisoner was found guilty as charged and ordered to repent his defiance of a papal order. He was allowed time to 'consider his position' and reluctantly, he confessed that he had gone too far. The story that, as he left the court, he muttered, 'But it does move,' is apocryphal. He was careful to show no defiance and it seems that his judges were reluctant to unleash the full fury of the Church's law against the aged celebrity. Three of them refused to sign the indictment. Initially sentenced to perpetual imprisonment, he was allowed to exchange this for house arrest in his own home.

A recent theory has suggested that the underlying reason for Galileo's condemnation concerned a much more sensitive theological issue. Like several other natural philosophers, he espoused atomism, the theory that all matter is composed of infinitesimal particles. This was regarded by some churchmen as being at odds with transubstantiation, the doctrine which explained the transformation of bread and wine into the body and blood of Christ. When Galileo's acceptance of atomism became obvious to readers of *The Assayer* it became another stick to beat him with – or so it is suggested.

Though it is quite likely that Galileo's enemies were ready to úse anything to achieve their ends, there was never any mention of eucharistic heresy in the records of the proceedings against him.

Galileo had never challenged any Church doctrine and always regarded himself as a loyal Catholic. He placed both his daughters in a convent of Poor Clares and, although there were economic reasons for this, one at least of the girls was both very devout and remained particularly close to her father. She took the religious name of Marie Celeste, in honour of the Virgin and of Galileo's work as an astronomer. For his part he was a practical friend of his daughters' convent and helped to fund various necessary repairs and improvements to the buildings. It was Marie Celeste's death, a few months after his ordeal in Rome, that broke him more than the decision of the Inquisition. What angered him was prejudice, the refusal of his critics to respond to or react against his discoveries with open minds. Whether it was slavish Aristotelianism or unyielding biblical literalism that set up barriers, he was determined to make breaches in them. But even he, with his monumental self-belief, failed to appreciate the force of the challenge thrown down to accepted 'truth' by the 'Dutch trunk'.

The difficulties experienced by Church leaders (or, at least, by the more intelligent among them) were not with Galileo's discoveries, but with their own foundation document, the Bible. There were devout and learned members of the ecclesiastical establishment facing up to this at the very time that Galileo was getting into trouble. One of them was Paolo Foscarini, Father Provincial of the Carmelite order in Calabria and Professor of Theology and Philosophy in the Jesuit University of Messina. In a work offered for publication in 1615 he stated the problem very simply:

> Because the common system of the world devised by Ptolemy has hitherto satisfied none of the learned ... a suspicion is

> risen up amongst all, even Ptolemy's followers themselves, that there must be some other system which is more true ... The telescope ... has been found out, by help of which many remarkable things in the heavens ... were discovered ... By this same instrument it appears very probable that Venus and Mercury do not move properly about the Earth, but rather about the sun ... Now can a better or more commodious hypothesis be devised than this of Copernicus? For this cause many modern authors are induced to approve of and follow it: but with much hesitancy and fear, in regard that it seemeth ... so to contradict the holy scriptures, as that it cannot possibly be reconciled to them. Which is the reason that this opinion has been long suppressed and is now entertained by men in a modest manner and as it were with a veiled face. *

Cardinal Bellarmine was no less frank in advising Foscarini not to pursue the matter at that particular time:

> If there was a real demonstration that the sun is in the centre and that the Earth goes round it, then one would have to proceed with much care in expounding the places of Scripture which seem to be contrary to that and it would be better to say that we do not understand them than to declare that false which has been demonstrated.†

* P. Foscarini, 'Epistle concerning the Pythagorian and Copernican Opinion of the Mobility of the Earth and Stability of the Sun', in *Mathematical Collections and Translations*, trs. T. Salusbury, 1661, p. 471

† Cf. F. J. Creham, 'The Bible in the Roman Catholic Church from Trent to the Present Day', in S. L. Greenslade, ed., *The Cambridge History of the Bible – The West from the Reformation to the Present Day*, CUP, Cambridge, 1963, p. 225

Foscarini's book was banned and he died the following year. One wonders whether, had he lived, he would have stood by Galileo when he faced his accusers.

What was at stake in the 'What goes round what?' argument was not the Bible, but Rome's monopoly of biblical interpretation. After all, it was not as though students of the Christian writings had only just realised that Joshua 10 and other passages were at odds with normal experience. It did not matter whether the Earth went round the sun or vice versa. What the Old Testament chronicle described was an interference with the laws of nature – i.e. a miracle. Neither Galileo nor any other contemporary scientist denied that God could perform miracles, so to interpret the heliocentric controversy as the opening salvo of the science versus religion war is a mistake. The Vatican's overreaction has to be understood in terms of major shifts in the tectonic plates of European culture. It was exactly a hundred years since the Pope's agents had made a similar ill-advised knee-jerk reaction to the protest of Martin Luther and a strong chord links the two events.

CHAPTER SEVEN

Confusion

The re-evaluation of every aspect of life – religion, morality, political organisation, social cohesion, family and neighbourhood relations – that flowed from the availability of vernacular Bibles resulted in dislocation and warfare throughout Europe that lasted into the eighteenth century. This was conflict of a kind the Continent had never known before.

Territorial aggression and dynastic rivalry (with which everyone was familiar) certainly played their part but now they were ennobled – or justified or excused – by ideology. Christians of various stamps took up arms against each other in the name of God. This violence was the backdrop against which philosophers, scientists, astrologers, prophets and all knowledge-seekers had to conduct their debates. Therefore, we must attempt a description – albeit sketchy – of the major conflict.

France's intermittent wars of religion began in 1562 and lasted well into the next century. The phenomenal growth of Calvinism was a serious concern of the government and the papacy. The Council of Trent was still in session when scores of Protestants (known as Huguenots) were massacred at Vassy on the orders of the Duke of Guise. There followed decades of civil war between religious groups led by rival aristocratic families. Assassinations and massacres received the backing of the Vatican but, despite a death toll of between two and four million, 'heresy' was not eradicated. Toleration of religious minorities was established by the Edict of Nantes in 1598 but this was only regarded by the government as a holding operation. Fresh restrictions were placed on Huguenots by the Peace of Alais in 1629 and in 1685 Louis XIV declared Protestantism once more to be illegal.

The continuing war in the largely Calvinist states of the Low Countries had resulted in atrocities such as the 'Spanish Fury' of 1576, when seven thousand citizens of Antwerp were slaughtered by mutineering Spanish troops. Such acts deeply scarred public memories and produced a bitterness that lasted many generations. In England, it was Catholics who found themselves to be a persecuted minority, suffering legal restraints for not supporting a state Anglican church that had a Calvinist theology and a semi-Catholic ritual. The government successfully fought off internal rebellions and foreign threats. For generations afterwards the nation celebrated annually the defeat of the Spanish Armada (1588) and the foiling of the Gunpowder Plot (1605).

The religious settlement did not satisfy everyone and when Charles I attempted, with the aid of the Anglican hierarchy, to establish autocratic rule he was opposed by a Puritan-led revolt, a major demand being for greater freedom of worship. Civil war raged from 1642 to 1651 and drew Scotland and Ireland into the conflict. Charles himself was executed in 1649 and republican government was established until the restoration of the monarchy in 1660. One feature of this disturbed period was the emergence of numerous religious and politico-religious fringe groups. All were firmly suppressed after 1660 and all Protestant 'non-conformists' joined Catholics as internal outlaws with few civic liberties.

In central Europe, Emperor Ferdinand II took it upon himself to stage a showdown with the Protestant rulers of his patchwork empire. The resulting Thirty Years War was the bloodiest conflict in the history of the continent prior to the twentieth century. The original combatants called upon co-religionist states for support. France, the Dutch Republic, England, Denmark and Sweden were drawn in, some by putting armies in the field, others by supplying mercenaries or financial aid. State treasuries were bankrupted. Acts of barbarity were witnessed, the like of which have seldom been seen in the annals of warfare. The beautiful city of Magdeburg was reduced to barren, smouldering ash in which most of its twenty thousand citizens perished and this was only the worst of several atrocities from which Europe took decades to recover.

It is scarcely surprising that, in an age when most people saw the hand of God in all major events, preachers and prophets proclaimed these disasters as Armageddon, the battle ushering in the end of the world. As for philosophers and men of science, their thinking was inevitably coloured by the tumultuous times through which they lived:

> In these years the most unheard of speculations were put forward, verbally and in print – demands for social and political equality, for a wide extension of the franchise, for abolition of a state church, for far-reaching social and legal reforms, for communism. All traditional institutions were called in question, including the Bible, private property, marriage and the family, male superiority. Where did all these fermenting ideas come from? *

In suggesting some answers to that question addressed by the great historian of seventeenth-century Britain, Christopher Hill, we can start by observing that not all the phenomena he was addressing were new. Rejection of authority had cropped up in peasant revolts and minor outbreaks of unrest often enough over the preceding centuries. The Reformation had offered new religious bases from which church and church/state authority could be attacked. Often associated with such expressions of dissent were radical strains of Christianity, prophecy and apocalyptic preaching. Existing tensions always tended to reach breaking point at times of civil unrest. It is

* C. Hill, *Some Intellectual Consequences of the English Revolution*, University of Winsconsin Press, Winsconsin, 1980, p. 7

not, therefore, surprising that new expressions of belief and unbelief should appear in the middle decades of the seventeenth century. Some of those expressions were simply manifestations of old scepticism and resentment. Others offered novel solutions to problems that had existed for generations.

Philosophers suggested fresh explanations of the relations between God, man, nature, church and state but they were not the only ones doing so. The vast majority of Europeans were not living on intellectual mountaintops and breathing the rarified atmosphere of rationalism in this turbulent, sometimes frightening, frequently confusing, epoch. They did not enjoy the luxury of leisurely speculation; they had to get on with the business of living – and of fitting their lives into a framework of beliefs that gave them meaning. In this chapter we shall try to discover what 'ordinary' people believed, and by 'ordinary' I mean men and women of all social classes and occupations who did not contribute to the academic debate about 'life, the universe and everything'.

For most seventeenth-century Europeans, religion was an amalgam of Christianity and folklore. God and the devil were invisible realities, as were fairies, goblins, demons and the saints. They manifested themselves through objects – holy relics, charms, herbal remedies and horoscopes – and through people – preachers, witches and magicians. The Reformation had removed some of these things from the spiritual landscape – but not all of them. And it had added other items to the environment – prophecy (especially the apocalyptic sort), personal interpretation of the Bible and divine endorsement of political change. It was the interaction of such forces that contributed not only to crises at the village level, but to the major political upheavals of the age that destroyed towns and decimated populations.

Common beliefs mingling ancient paganism, old Catholicism and new Protestantism continued throughout the century and beyond. It was not 'science' that cut through the confusion. It was industrialisation after *c.* 1750 that, by creating urban sprawl and depopulating rural areas, brought about a new kind of society (with problems of its own). Even in the early nineteenth century there were enough folk legends and traditional stories available to make up the collections of the Brothers Grimm.

Let us start by examining witchcraft and the war against it. Statistics are notoriously unreliable in any analysis of such social phenomena but if we want to understand how widespread this one was we can take as a rough guide some two sets of figures that have been compiled by historians. In a corner of south-west Germany bordered (today) by Austria, Switzerland and the Czech Republic it has been calculated that, in the century following 1560, more than 3,200 people were executed as witches. Throughout a similar period in south-eastern England (Essex, Hertfordshire, Kent, Surrey, Sussex) about one hundred people were found guilty under the prevailing anti-witchcraft laws and hanged.

Persecution was, of course, sporadic. A witch-hunt could be set off by any of several events – an outbreak of cattle disease, the arrival of a fanatical preacher or (and we should never exclude this from our investigation) the activities of a practitioner boldly claiming to exercise demonic power. And we must never lose sight of the major upheavals in society to which we have already referred. As the Four Horsemen of the Apocalypse – War, Pestilence, Famine and Death – galloped over the land, sufferers, in their misery, often looked for scapegoats.

To take just one example of an outbreak of rage and hate, in one small German rural locale with a population of some seven hundred people, records show the execution of fifty-four victims in a three-year frenzy of witch mania.* If the total statistic is made up of many

* Cf. N. Cohn, 'The Non-Existent Society of Witches', in *Witchcraft and Sorcery*, ed. M. Marwick, Penguin, London, 1986, p. 155

such instances of spontaneous persecution, we may reasonably infer that there were large tracts of country and several periods of time relatively unaffected by militant expressions of fear and hate. In fact, three quarters of Europe did not witness a single witch trial.

There was, however, a widespread and quite detailed folklore about sorcery that fed a common appetite for the arcane (and, lest we should dismiss such fascination as the ignorance of unsophisticated, credulous people, let us not close our eyes to the phenomenal success in our own day of the *Harry Potter* films and other supernatural fantasies that grace our large and small screens). Various versions of the 'sabat' legend were in circulation. These referred to gatherings of witches and wizards believed to be summoned by Satan, usually in the form of a monstrous beast. Attendant sorcerers travelled long distances through the air to be present at these rituals in which all manner of obscene and profane acts took place. The mass and other Christian ceremonies were parodied. The devil's disciples were made to repeat their oaths against God and his followers and swear to thwart the work of the Church, both Catholic and Protestant.

Various attempts have been made to trace such legends back into Europe's pre-Christian, pagan past, old fertility cults and even Greco-Roman mystery religions. Whatever similarities may exist are probably the result of that consciousness of the conflict between good and evil that is shared by all cultures. What matters for our purposes is that they had become firmly embedded in the common psyche of seventeenth-century European men and women. Church leaders of any persuasion had little incentive to dislodge them. Shared dogma insisted that a conflict between good and evil existed in the heavenly realm and was constantly being played out on the stage of human history. The New Testament was quite clear about this. St Paul had exhorted believers to stand firm against the devil's 'evil tricks'. 'We wrestle not against flesh and blood,' he insisted, 'but against spiritual wickedness in the heavenly places.'*

We have seen how, in the Middle Ages, this conflict had been luridly depicted on church walls throughout Europe by artists who portrayed the denizens of hell as hideously as their imaginations would allow. The Protestants' stripping of 'superstitious' imagery from churches may have removed such terrifying permanent reminders of 'the devil and all his works' from most church interiors but the message was thereafter conveyed in other equally sensational – and portable – forms, thanks to the printing press. Pamphlets, ballads and broadsheets with vivid illustrations were available in profusion. They told stories of men or women who had made pacts with the devil, performed appalling crimes with the powers he had provided but had, ultimately, paid the price for their folly.

One of the best known of these tales, because it became the subject of a stage play by Thomas Dekker, William Rowley and John Ford, was *The Witch of Edmonton*, first performed in 1621. The plot concerns Elizabeth Sawyer, a poor, elderly woman who makes a pact with the devil in order to be revenged on her neighbours for their sleights and cruelties. She thus becomes the stereotypical witch. She drives one of her tormentors to suicide, is discovered and is hanged. There are subplots in the play involving characters who are, in some ways, no better than the witch, so that, intriguingly, the audience was left with an enigma. This is interesting because it is one intimation that attitudes towards those hounded as witches were far from black and white. We will return to that shortly. First we must consider what has been called the 'great witch hunt'.†

Although outbreaks of activity, as we have said, were sporadic, they were sufficiently frequent and similar for us reasonably to

* Ephesians 6, 11:12

† N. Cohn, op. cit., pp. 154f.]

identify witch mania as a feature of life throughout the period 1600–1680. In Catholic lands the *Malleus Malificarum* went through sixteen reprints between 1574 and 1669 and was influential in many, if not most, prosecutions.

The tendency was for persecution to erupt from small beginnings. Misfortune would strike an individual or group who, seeking a scapegoat or a release of their anger, would accuse an unpopular member of the community of *maleficium* (malevolent sorcery). The complaint would be taken up by the Church authorities, either because they did not want to be seen as complacent or because they had their own reasons for launching a general purge. Suspects investigated, usually under torture, were pressed to indict others as members of their covens and, before the local people knew what had hit them, the wildfire of hatred and fear was raging all around them.

Southern Germany, in the areas centred on Bamberg and Würzburg, saw the worst atrocities. The outbreak in Bamberg in 1526 came in the wake of a devastating winter that had frozen the crops. The prince bishop enthusiastically took up the prosecution that resulted in so many arrests that a special prison had to be built. Between three hundred and six hundred men, women and children were executed. Witch-mania spread throughout the region. The word 'holocaust' is not too strong for the horrors witnessed in the period up to 1631. The following catalogue of fatalities is not and cannot be complete because accurate records have not survived. Even so, they tell a dismal story:

Würzburg and environs	600
Mainz	650
Eichstatt	274
Ortenau	70
Reichertshofen	50

Offenburg	79
Coblenz	24

Adding in the burnings (most witches were burned) in smaller centres and rural areas and allowing for incomplete records brings the total to something well in excess of three thousand.

Statistics, of course, do not convey either the horror or the complexity of the events of those few years. Interrogation and the networks of informers resulted in individuals of all ages and all ranks being entrapped. One contemporary reported on the situation in Bonn:

> There must be half the city implicated for already professors, law students, pastors, canons, vicars and monks have been arrested and burned. His princely grace [the Prince-Archbishop of Trier] has seventy wards who are to become pastors, one of whom, eminent as a musician, was yesterday arrested. Two others were sought for but have fled. The chancellor and his wife and the private secretary's wife are already executed. On the eve of Our Lady's Day there was executed here a maiden of nineteen who bore the name of being the fairest and most blameless of all the city and who from her childhood had been brought up by the bishop himself. A canon of the cathedral, named Rotenhahn, I saw beheaded and burned. Children of three and four years [are said to] have devils for their paramours. Students and boys of noble birth of nine, ten, eleven, twelve, thirteen, fourteen years have been burned. In fine, things are in such a pitiful state that one does not know with what people one may talk and associate.*

* G. L. Burr, ed., 'The Witch Persecutions', in *Translations and Reprints from the Original Sources of European History*, University of Pennsylvania Press, Philadelphia, 1898–1912, III, 4, pp. 18–19

We cannot even guess at the jealousies, rivalries and class hatreds that lay beneath all this suffering. We can, however, indicate one conflict closely related to the witch craze: the Thirty Years' War. Catholic rhetoric labelled witchcraft as a heresy (or, perhaps, labelled heresy as a form of demonic activity). The ecclesiastical powers used ruthless imposition of anti-witchcraft legislation as a means of reinforcing their authority and warning those tempted to freethinking or Protestant sympathies of the inviolability of Catholic truth. In some cases, the only escape from persecution came when the tide of war turned and Protestant victors swept into town. In Bamberg, whose tragic story was similar to Würzburg's and where another thousand supposed sorcerers perished, the reign of terror came to an end with the arrival of a Swedish army. They opened the prisons and forced the prince-bishop to flee.

Catholic authorities were responsible for the great majority of prosecutions but that does not mean that trials did not take place in Protestant lands. Zealous Calvinists and Lutherans were just as appalled as their Catholic opponents by tales of sabats, curses and spells. One reason for the discrepancy is the different legal systems operating in various countries. Where Roman law prevailed, or where Church courts enjoyed independence, prisoners could be prosecuted for diabolism. Lands such as England and Sweden had common law codes which only dealt in *maleficium*, offences against persons or property.

> The forms, wherein Satan [obliges] himself to the greatest of the magicians, are wonderful curious; so are the effects correspondent unto the same: For he will oblige himself to teach them arts and sciences, which he may easily do, being so learned a knave as he is: To carry them news from any part of the world, which the agilities of a spirit may easily perform: to reveal to them the secrets of any persons, so

> being they be once spoken . . . Suchlike, he will guard his scholars with fair armies of horsemen and footmen in appearance, castles and forts, which all are but impressions in the air, easily gathered by a spirit, drawing so near to that substance himself.*

These words from an earnest treatise by James VI of Scotland (shortly to be James I of England) indicate that belief in demonic powers was not confined to simple people. The prince of darkness and his minions were as familiar in the palaces of kings as in the hovels of peasants.

> From Paris there is a strange tale that the French king hath been miraculously delivered. His Majesty was about to partake of the communion upon the day of Corpus Christi and had opened his mouth to receive the Host when suddenly there appeared a dog which pulled him backwards. A second time the king essayed to receive the Host and again the dog prevented him. Hereupon he commanded the priest to partake of the Host himself, which at first he would have refused, had not the king importuned him. When the priest had taken it, he swelled up and his body was burst in twain. Thus was the plot discovered and some of the noblemen who were privy to it are now in the Bastille.†

So ran a rumour current in London in 1604. The origin of the story or the extent to which it had been embellished in the retelling is not

* James IV of Scotland, *Daemonologie, In Forme of a Dialogue*, 1603 ed., p. 21

† G. B. Harrison, *A Jacobean Journal – Being a Record of Those Things Most Talked About During the Years 1603–1606*, Routledge, Oxford, 1941, p. 152

to the point. What is significant is that it was readily believed by the new English king and his people.

James had absorbed from his Presbyterian upbringing a firm belief in the activities of the devil and for him it became so important a subject that he felt it necessary to venture into print himself in order to warn his subjects to be on the alert. Only a few years previously he had been involved in Scotland's most systematic witch-hunt. In 1589 James' betrothed bride, Anne of Denmark, set sail from Copenhagen for her new country but her convoy was immediately battered by a ferocious storm that forced it to take refuge in Norway. Since, as everyone knew, such hazards had spiritual causes, search was made for Satan's human agents.

Several prominent people with connections to the Danish court were indicted for consorting with demons to drown Princess Anne and two self-confessed witches were burned. These events had a profound effect on Anne's twenty-three-year-old bridegroom. When, two years later, he heard of a coven in the environs of North Berwick accused of raising storms and other sorceries he took a close interest and attended some of the interrogations. He learned how the devil in person had hosted a sabat in North Berwick church, wearing a tall black hat and preaching from the pulpit. The investigation pursued its by now conventional form and, within weeks, over a hundred suspects had been arrested on various charges including attempting to poison the king and sink his ship. These high-profile cases kept alive an anti-witch frenzy that, over the following century, resulted in hundreds – perhaps thousands – of deaths.

The most notorious of England's Protestant 'inquisitors' was Matthew Hopkins, the self-styled 'Witch-Finder General'. Much has been made of the exploits of this zealous young Puritan (he was only twenty-seven when he died). There is no doubt that he and his assistants were responsible for the deaths of as many as three hundred convicted witches in East Anglia and neighbouring counties between 1644 and 1647 but under no circumstances can this persecutor be regarded as a 'typical' product of Puritan oppression. No record exists of any official appointment by parliament. Virtually nothing is known about him apart from his one-man crusade and his very notoriety makes him a special case. During his brief career a number of local justices expressed concern at Hopkins' methods and questioned his motives. This 'scourge of the devil' was, himself, a sick man who was obliged to give up his brief career in 1647 and who died, probably of pleurisy, a short while afterwards. We would probably be close to the truth if we regarded Hopkins as one example of the tensions and animosities in a nation convulsed by civil war.

Throughout the seventeenth century, scepticism struggled with belief, only growing stronger as a measure of peace was restored to Europe in the latter decades. In 1604 the English law against witchcraft was actually strengthened. Hitherto it had been concerned primarily with *maleficium*. Now it became an offence to 'consult, covenant with, entertain, employ, feed or reward any evil and wicked spirit to or for any interest or purpose'. Around the middle of the century Robert Boyle, the pioneer chemist, physicist and inventor, declared himself a convert – not *from* belief in devil worship but *to* it. He was convinced by the numbers of cases of successful prosecutions in several countries. He was not alone among men of science who shared the common belief that satanic forces existed and that some men and women were in league with them.

Paradoxically, it was the very success of witchcraft prosecutions and the thousands of burnings and hangings that helped to turn the tide. When decent, law-abiding Christian citizens went in fear of the knock on the door they began to question not just the activities of the witchfinders, but the theology that lay behind their persecution. Orthodox believers had to take care when unravelling myth and malice from basic Christian truth: were self-confessed sorcerers simply deluded? Was it possible to make a pact with Lucifer? Did Lucifer really exist? And, if not Lucifer, what of God? The twisting path of scepticism could – and in some cases did – lead to the precipice of atheism.

As early as the 1590s the confusion existing in many minds was evidenced in Marlowe's *Doctor Faustus*. In conversation between the doctor and the demon Mephistopheles, the following exchange occurred:

> Faust: Come, I think hell's a fable.
> Mephistopheles: Ay, think so, till experience change thy mind.
> Faust: Why, think'st thou, then, that Faustus shall be damn'd?
> Mephistopheles: Ay, of necessity, for here's the scroll wherein thou have given thy soul to Lucifer.
> Faust: Ay, and body too: but what of that?
> Think'st thou that Faustus is so fond to imagine
> That, after this life, there is any pain?
> Tush, these are trifles and mere old wives' tales
> Mephistopheles: But, Faustus, I am an instance to prove the contrary:
> For I am damned and am now in hell.
> Faust: How? Now in hell? Nay, and this be hell, I'll willingly be damn'd here.
>
> Christopher Marlowe, *Doctor Faustus*, *c.* 1592, Scene 5

Senior churchmen, such as Samuel Harsnett (later Archbishop of York) distanced the Church from the crusade against witches:

> They that have their brains baited and their fancies distempered with the imaginations and apprehensions of Witches,

> Conjurors, and Fairies and all that Lymphatical Chimera, I find to be marshalled in one of these five ranks: Children, Fools, Women, Cowards, sick or black melancholic discomposed wits.*

By the end of the seventeenth century the witch-hunt conflagration had been reduced to a few smoking embers. In 1684 the last witch-hanging occurred in England. By 1712 the situation can be well illustrated by the conviction of the Hertfordshire 'witch' Jane Wenham and the aftermath. Jane's case was typical: an unpopular old lady denounced by some of her neighbours for doing them various mischiefs. The prosecution was taken up eagerly by local clergy. Wenham was tried at Hertford before fair-minded and commonsensical Sir John Powell. He tried to steer the jury towards an acquittal. For example, he told them to ignore evidence that Jane was able to fly because 'there is no law against flying'. When, despite his efforts, the jury returned a guilty verdict, he had no alternative but to sentence the defendant to death. But he immediately appealed to the Crown for a pardon and Jane was subsequently released.

This unleashed a mini-pamphlet war. The first to leap into the breach was the anonymous writer of *A Letter from a Physician in Hertfordshire to his Friend in London*. He opened by indicating the consideration that had so often silenced critics:

> I am fully aware to what hazards a man of public character exposes his reputation to in . . . writing on such a topic, especially in the country where to make the least doubt is a badge of infidelity and not to be superstitious passes for a dull neutrality in religion, if not a direct atheism.

* S. Harsnett, *A Declaration of Egregious Popish Impostures*, 1603, p. 131

The author's target is the clergy whom he accuses of both pride and ignorance:

> If . . . the clergy would be a little more conversant with the history of diseases and enquire more narrowly into the physical causes of things, several effects would not appear so perplexing, neither would they be so forward as to ascribe those diseases to the devil, when nature is primarily concerned.*

Such attacks provoked a flurry of literary responses. In the atmosphere of free debate that now prevailed, arguments for and against demonology could be openly rehearsed and it was at this time that the science versus religion contest began to take shape. The Hertfordshire doctor drew attention to the foundation of learned societies, which he saw as indicators of a brave, new utopia:

> Not only our witchcrafts have been banished, but all arts and sciences have been greatly improved. Our buildings are much more beautiful and commodious and yet more cheaply built and easier kept in repair. Our gardens and orchards are stocked with new and nobler fruits and fields and woods with useful trees. Many of our lands that were almost useless are loaded with new kinds of grass and roots, by better understanding the improvement of the soil. Our money is more beautiful and less liable to being impaired. Physic and surgery are new moulded and improved, for the lengthening out of life in ease. The smallest parts of bodies are made viable by glasses and the farthest planets are brought near and their motions wonderfully accounted for. Navigation is much improved, and

* *A Full Confutation of Witchcraft, More Particularly of the Depositions Against Jane Wenham, Lately Condemned for a Witch at Hertford*, 1712, pp. 3–4, 48

> communications of knowledge settled with the farthest parts. All arts are improved, God is seen and admired in his works, and the honour of religion no ways lessened.*

Clergy who had kept a low profile during the years of Puritan domination or who had actually taken part in the witch trials were now keen to distance themselves from what were increasingly regarded as cruel and non-biblical activities. In 1718, Francis Hutchinson, Bishop of Down and Connor, offered a rather lame apology for his brethren involved in the Wenham case in *An Historical Essay Concerning Witchcraft – with Observations Upon Matters of Fact, Tending to Clear the Texts of the Sacred Scriptures and Confute the Vulgar Errors.* In his dedication he lamented the suffering of all who had been condemned before and wrote, 'I humbly offer my book as an argument on behalf of all such miserable people who may ever in time to come be drawn into the same danger in our nation.' At least Hutchinson could soothe his conscience with the knowledge that no witch had been hanged in England for more than thirty years and that things were worse elsewhere. The last witch execution did not take place in Scotland until 1727, France 1745, Germany 1775 and Switzerland 1782.

Most thinking people who seek answers to the problems of 'life, the universe and everything' want two things – certainty and freedom. They want to know that their search for meaning has led them to ultimate truth and that they have liberty to seek, formulate and live by that truth. For atheists the result of their investigation is very satisfactory: since no ultimate source of ethical authority exists they are free to live their lives exactly as they wish, choosing for themselves whatever moral restraints (if any) they elect to impose upon themselves. For the rest of humanity life is not so simple.

* Ibid., p. 170

The existence of a supreme being, one of whose functions is that of lawgiver, imposes a framework within which they are constrained to live. For seventeenth-century European believers there were three authoritative channels through which divine instructions were relayed – Church leaders, the Bible and the inner promptings of the Holy Spirit. As, first, the papal hierarchy and, subsequently, the leadership of Lutheran, Calvinist, Anglican and Zwinglian communions was called in question, old Christendom fragmented into a profusion of churches, sects and cults. It is not part of my remit to peer into the kaleidoscope of Anabaptist, quietist, millenarian, familist and other religious and politico-religious groups whose clashing shapes and colours made up society in that troubled era. There is, however, one feature that is particularly pertinent to this study. It was an exclusive phenomenon that was owned by most religious communities but which also exerted an immense influence over society as a whole. Prophecy.

In a confused and troubled age people wanted, more than ever, to know what the future held in store. Everyone, educated and uneducated alike, resorted to foretellers of one kind or another. Basically, there were four sources of information about future events: judicial astrology (which we have already encountered – see pp. 13, 64), folklore, biblical exposition, and independent soothsayers.

The advent of printing had made astrological predictions available to a much wider audience by introducing a new literary genre, the almanac. These annual tables containing astronomical information, important dates and astrological predictions, appeared in profusion in every country as pocket books and they were immensely popular. In some lands they outsold the Bible. Now, without going to the expense of having a horoscope prepared, anyone could read his/her destiny in the stars. It is estimated that, around 1650, some four hundred thousand copies of various almanacs were being sold annually in England. Their quality and usefulness varied enormously.

Some were published by experts such as Tycho Brahe and Kepler. Others vied for market share by offering sensationalist predictions.

That belief in the influence of the planets was widespread and powerful is demonstrated by the phenomenon of Black Monday, 29 March 1652. John Evelyn noted in his diary that it 'so exceedingly alarmed the whole nation, so as hardly any would work, nor stir out of their houses'.* He was not exaggerating. Throughout the country there were scenes of great alarm and distress. Simple people in Dalkeith lay on the ground wailing and entreating Christ to have mercy on them. Sober London merchants piled their treasures into their coaches and fled the city. The reason? An eclipse of the sun – or rather the wild almanac prognostications attached to this rare event. Some had forecast the downfall of governments, the death of kings and the onset of mob rule.

Inevitably, at a time of civil unrest and revolution, almanacs and pseudo-astrological publications were used for political purposes. The most celebrated contributor to the Black Monday panic was William Lilly, friend of politicians like Bulstrode Whitelocke, Lord Keeper of the Great Seal, and scholars of the stature of Elias Ashmole, founder of the Ashmolean museum, Oxford. He was a protagonist in the 'war of the almanacs', using his publications to counter those of the royalist astrologer, George Wharton. Such soothsayers were just as influential (if not more so) than the preachers who accompanied the armies of King and Parliament in the British Civil War.

Lilly was the foremost, in his day, in championing the prophecies of Merlin Ambrosius, recorded in Geoffrey of Monmouth's twelfth-century *Historia Regum Britanniae*. This list of obscure 'revelations' about forthcoming events had long been a source of material for almanac writers and others hoping to influence contemporary events by providing them with the imprimatur of the legendary Arthurian

* *The Diary of John Evelyn*, E. S. de Beer, ed., OUP, Oxford, 1959, p. 319

wizard. This is an example of how Lilly applied Geoffrey's text to the fate of the Stuart kings:

> He calls King James, the Lion of Righteousness and saith, when he died . . . there would reign a noble White King. This was Charles the First. The prophet discovers all his troubles, his flying up and down, his imprisonment, his death and calls him Aquila. What concerns Charles the Second, is the subject of our discourse.
>
> 'After them shall come through the south with the sun, on horse of trees and upon all waves of the sea, the Chicken of the Eagle, sailing into Britain and arriving across to the house of the Eagle, he shall show fellowship to these beasts.
>
> 'After the Chicken of the Eagle shall nestle in the highest rock of all Britain: nay, he shall nought be slain young; nay, he shall nought come old . . .

THE VERIFICATION

'His Majesty being in the Low Countries when the Lord General had restored the secluded Members, the Parliament sent part of the Royal Navy to bring him for England, which they did in May 1660. Holland is East from England, so he came with the sun; but he landed at Dover, a part in the south part of England. Wooden-horses are the English ships . . .

'The Lord General and most of the gentry in England, met him in Kent and brought him unto London, then to Whitehall.

'Hereby the highest . . . Rock is intended London, being the metropolis of England.'*

* W. Lilly and E. Ashmole, *William Lilly's History of his Life and Times From the Year 1602 to 1681*, Harvard University Press, Massachusetts, 1822 ed., pp. 194–6

By no means everyone swallowed such oracles. Several writers lampooned the Merlin legends and those who gave them credence. In his *British History*, Milton dismissed Geoffrey of Monmouth's 'fabulous book' in its entirety, remarking how strange it was that it had been 'utterly unknown to the world, till more than 600 years after the days of Arthur'.*

The significant point is that it *was* at this particular time in the nation's history that people were turning to ancient prophecies and any writings that offered to make sense of the trauma Britain and Europe were going through. Merlin's supposed predictions were not the only ones to be widely proclaimed. It was in 1641 that the prophecies of Mother Shipton, a soothsayer who lived (*c.* 1488–1561) in Knaresborough in the mid-sixteenth century, were published over and again in various versions. A witness of the Great Fire of London in 1666, writing to his friend Viscount Scudamore, reported that thousands believed that the tragedy had been foretold by Mother Shipton and that many citizens were reluctant to attack the blaze since the devastation was predetermined.†

Contemporary with Mother Shipton was the more famous Nostradamus or Michel de Nostredame, a French apothecary and, reputedly, judicial astrologer (a title he rejected). Unlike Shipton, Nostradamus was a considerable scholar who drew on his knowledge of classical and medieval literature in forming his analysis of contemporary and future events. Since most of his prophecies concerned disasters of various kinds it is hardly surprising that in an era replete with disasters people should turn to Nostradamus' obscurely worded oracles in an effort to understand what was happening to them. It is

---

* Cf. R. F. Brinkley, *Arthurian Legend in the Seventeenth Century*, Routledge, Oxford, 2016, p. 81

† S. G. Bell, *The Great Fire of London in 1666*, Bodley Head, London, 1923, p. 316

not difficult to see how readers scanning the obscure quatrains would have readily identified what such words as these presaged:

> The blood of the just will be demanded of London.
> Burned by fire in the year '66.
> The ancient lady will fall from her high place
> And many of the same sect will be killed.

One of Nostradamus' sources was the Bible. Preachers and humble readers routinely turned to its pages for enlightenment, particularly about portents of the 'Last Days'. The books of Daniel and Revelation contain prophecies relating to the end of human history that are both detailed and wrapped in poetic language. Probably there has never been a time in Christian history when expositors have not pored over these texts and attempted to identify their signs and portents in terms of their own contemporary events. But few, if any, eras have seen more frenzied apocalyptic predictions than the seventeenth century in Europe.

Richard Farnham, a Colchester weaver, was just one of the wild spirits produced by this wild age. Arrested in 1636 on the orders of the ecclesiastical court, he declared to his accusers that he had been assigned by God to issue dire warnings of imminent disaster: 'I am one of the two witnesses that are spoken of in the eleventh chapter of Revelation . . . The Lord hath given me power for the opening and shutting of the heavens.' His rejection of mere human convention permitted him to take to wife a woman who was already married. He spent most of the next five years in prison and died of plague early in 1642 but his followers ardently believed that he had been resurrected and transferred to another land to continue his mission.*

* Cf. Keith Thomas, op. cit., p. 159

It was not just poor artisans who felt compelled to proclaim messages revealed to them from heaven. Lady Eleanor Douglas, a lady of breeding who enjoyed entrée to the royal court, was one. Between 1625 and 1652 she issued numerous prophecies, including the accurate prediction that Charles I's favourite, the Duke of Buckingham, would be assassinated. Various bouts of imprisonment and incarceration in the Bethlem Royal Hospital for the insane could not silence her because her conviction that Judgment Day was near at hand lent urgency to her mission.

One of the radicals briefly under Lady Douglas's wing was Gerrard Winstanley, leader of the True Levellers or, as they are better known, Diggers. They embraced a form of communism based on the pattern of the first Christians as recorded in Acts of the Apostles, chapter 2, in which Christ's disciples were described as selling all their possessions and distributing the proceeds to the poor. Winstanley set out his doctrine in a pamphlet, *The New Law of Righteousness,* presenting himself as the prophet of the coming divine order in which there would be no hierarchy and no private property. It goes without saying that the Diggers did not last long.

After four world empires have come and gone, 'The God of heaven will establish a kingdom that will never . . . but will completely destroy other empires and last for ever.'* This was the foundation text upon which the Fifth Monarchists, yet another sect thrown up by the Civil War, based their vision of the future. They supported Oliver Cromwell and saw in him the man who would establish the 'rule of the saints', destined to hold all in readiness for the return of Christ in 1666 (666 being, according to *Revelation,* the mystic number of the last enemy to be overthrown before the Apocalypse). The Fifth Monarchists were influential in shaping the new order after the execution of Charles I but later became disillusioned with Cromwell and faded from the

* Daniel, 2:39

limelight. In 1661 a group of them made a suicide raid on London in the name of King Jesus. Its only achievement was to make the repression of all unorthodox sects more severe.

Such examples (and one could cite several others who made the headlines with their speculative interpretation of the Bible) have often given rise to the view that seventeenth-century apocalyptic visions were the preserve of extremist fanatics teetering on the brink of insanity. Nothing could be further from the truth. The imminent return of Christ was part of the official theology of all mainstream churches. National leaders interpreted their role as preparing for this event. Cromwell certainly held to this conviction. His welcoming of Jews back into England after 361 years was largely motivated by another of the signs and portents mentioned in Scripture: the conversion of the Jews was a necessary precursor of the Second Coming.

And somewhere in the shadows of history were a few rare men – devout and methodical thinkers – who, when the clamour of war had ceased and the dust of broken cities had settled, would be recognised as prophets of a different kind. They would have an influence more lasting than the wild-eyed preachers and the desperate revolutionaries. One such was Joseph Mede (1586–1638), fellow of Christ's College, Cambridge. He was a thoroughly conventional Anglican who studiously avoided religious controversy. He also shunned offers of preferment and for much the same reason: he did not want to be diverted from his studies. The width and depth of his knowledge were prodigious. He lectured in Greek and had command of more than one Semitic language. He was a natural philosopher and a practising anatomist. His studies embraced mathematics, botany, physics, theology and Egyptology. He mixed with the leading English thinkers of the day and was in regular correspondence with several European scholars. He had the scholar's open yet critical mind. His guiding principle in scientific, as in theological, debate was, 'I cannot believe that truth can be prejudiced by the discovery of truth.'

By the time the politico-religious crisis fell upon his country Mede was long dead but he spoke into that crisis through his writings. Foremost among them was *Key of the Revelation Searched and Demonstrated*, published in Latin in 1627 and translated into English in 1643 by order of Parliament. It was a scholarly exposition of the last book of the Bible, the apocalyptic Revelation of St John, but it was not the kind of sensational prophesying that sought to equate biblical imagery with current events. Mede treated the text as something to be understood for *itself*. Working on the established principle of Protestant exegesis that the Bible is a coherent whole and all its parts, when properly understood, knit together, he went systematically through Revelation, cross-referencing it with other biblical material and came up with a chronological schema of the End Time. His verdict was that the portents referred to in Scripture had not yet all been fulfilled and that, therefore, the Second Coming was still a future event. Moreover, it could not be hastened by human agents attempting to 'force God's hand'. It was this balanced and painstaking millennialism that influenced a generation of Christian thinkers, including the German Samuel Hartlib, the Czech Comenius and, among his fellow countrymen, Milton, Locke and Newton.

CHAPTER EIGHT

I think – so what?

One man responded angrily to the treatment of Galileo in the 1630s. He vented his feelings in a letter to a friend:

> I have decided to fight . . . with their own weapons the people who confound Aristotle with the Bible and abuse the authority of the Church in order to vent their passions – I mean the people who had Galileo condemned . . . I am confident I can show that none of the tenets of their philosophy accords with Faith so well as my doctrines.*

The writer was René Descartes (1596–1650). He was, and remained, a Catholic, even though he rebelled against his upbringing, but he had more in common with Protestant philosophers and theologians, many of whom strove to present Christianity as a rational creed.

Before we take up Descartes' story we should remind ourselves yet again of the background to his life. The Thirty Years' War was devastating. The English Civil War cost the lives of a quarter of a million people – in terms of percentage population, a figure that far exceeds British losses in the First World War. The seventeenth century was a bloody era. The conflicts were not caused solely by religion but leaders on both sides certainly used religion to justify their aggression. In England Charles I demanded the loyalty of his subjects because he was God's anointed ruler. He was supported by a growing number of churchmen (Arminians) who were in theological conflict with

* 'Letter to Mersenne', 31 March 1641, cited in Z. Janowski, *Cartesian Theodicy: Descartes' Quest for Certitude*, Springer, New York, 2000, p. 156

Calvinism over the issue of predestination. This forced Puritans and other Calvinists (who were well represented in the House of Commons) to progress from a religious to a political radicalism.

Small wonder that thinking men asked themselves afresh the old question, 'What's life all about?' and challenged the answers given by their forefathers. Revelation tended to be disregarded, leaving reason as the only road to truth. Some – the rationalists – believed this to be a reliable route. Others – the sceptics – asserted either that there were no final answers or that, if there were, they were beyond the wit of man to discern. Meanwhile, away from the rarified atmosphere of philosophical debate, in what we might call the 'real world', established churches were failing to hold onto their congregations. Beggars, vagabonds and unemployed rabbles roamed the countryside. There was a general culture of hedonism and loose morality. 'Millenarian and chiliastic ideas were enjoying considerable appeal, as wandering prophets and prophetesses interpreted the signs of the times and diagnosed the fate of the world.'*

The pioneer of rationalism, René Descartes, was someone who had been brought up with Catholic certainties, had tested them on the battlefield and found them wanting. He was educated by Jesuits who gave him an excellent grounding in mathematics but did not satisfy his spiritual quest. At the outset of the Thirty Years' War he entered the army and saw service in Bohemia, Hungary and Holland. The horrors he witnessed did not undermine his faith. Rather they made him determined to find a more secure basis for it than the tired scholasticism still stubbornly adhered to by his teachers. It was while on active service that he had a dream or vision that moved him deeply. In it he 'discovered the foundations of a marvellous science and, at the same time, my vocation was revealed to me'. He settled in Antwerp, a refuge for many

* M. Fulbrook, *Piety and Politics – Religion and the Rise of Absolutism in England, Württemberg and Prussia*, CUP, Cambridge, 1983, p. 24

freethinkers, where, as a gentleman of independent means, he developed in a leisurely – not to say languid – fashion a comprehensive philosophy to unite all *scientia* in one gloriously foolproof system.

Descartes was something of an eccentric. Claiming that he did his best work in bed, he seldom arose before midday. He refrained from reading widely, claiming that he did not want his thinking to be shaped by other men's ideas. He made no attempt to seek academic preferment, believing that intellectual conflict would consume too much time and energy. He was no energetic controversialist, prepared to enter the cut and thrust of debate. Having just completed, in 1633, a treatise, *The World,* which endorsed heliocentrism, he hastily withdrew it from publication on hearing of Galileo's conviction. 'It is imprudent to lose one's life when one can save oneself without dishonour,' he somewhat lamely explained. Despite the belligerent tone of the 1641 letter quoted on page 173, Descartes was clearly on the defensive about his relationship with the Catholic Church. In another letter of 1644, he was at pains to claim that anyone who took the trouble to understand his position on heliocentric theory (still actually hidden from view in his unpublished manuscript) could not possibly censure him:

> . . . it is more necessary to say that the Earth moves, if one adopts the system of Tycho, than if one accepts the Copernican system . . . If one cannot accept either of these two systems, it would be necessary to go back to that of Ptolemy. I do not think the Church ever desires us to do that, since that system is manifestly contrary to experience. All the scriptural passages that are contrary to the Earth's motion have nothing to do with astronomy; they concern only a manner of speaking.*

* Cf. D. M. Clarke, *Descartes: A Biography*, CUP, Cambridge, 2006, p. 413

Descartes' position was essentially that taken by Rheticus (see p. 64) and most Copernican astronomers. We might call it the 'historical' explanation: what the Bible described was the truth as it appeared to writers with the knowledge available *at the time of* writing. But this still did not satisfy literalist Bible apologists and it was they who held the ring in Rome.

Descartes dabbled with, but eventually abandoned, medicine. He applied his fertile mind to several problems in the realms of pure and applied mathematics and wrote several works on geometry and philosophy during his long sojourn in the Netherlands, before accepting a post at the court of the equally unorthodox Queen Christina of Sweden. As Bertrand Russell tartly remarked, she was a lady who, because she was a monarch, thought that she had the right to waste the time of great men. In her service Descartes was obliged to change his daily regime, rising at five in the morning to teach his pupil who would only allocate that early hour to her lessons. It was probably this and the Scandinavian climate that accounted for his death from pneumonia at the age of fifty-three.

In Descartes we see the flowering of that individualism championed by both humanist and evangelical writers. He devised a way of thinking that would be based *entirely* on reason. However, his primal motivation could not discount emotion. As he confessed on more than one occasion, 'Some years ago I was struck by the large number of falsehoods I had accepted as true in my childhood . . . I realised that it was necessary to demolish everything . . . and start again right from the foundations'. He shared the anger and disillusionment felt by many with the battle of religious certainties that was destroying so many lives. Yet he, too, craved certainty and needed some sort of god. Reclining on his pillows, he concluded that every man can *think* his way to truth – indeed that this is the only road to tread. But thinking must be rigidly disciplined.

What the philosopher needed was a regime, a pattern to follow, hence his early works, *Rules for the Direction of the Mind* (1628) and *Discourse on Method* (1637). The first step along the rationalist road, he asserted, was to empty the mind of all preconceptions. This was achieved by *methodological scepticism*: one must doubt every idea until one has proved it. Into the void one must place only information whose reliability is evidenced by either intuition or deduction. By intuition he meant facts that were self-evident (healthy babies are born with ten fingers and ten toes; 2 + 2 = 4, etc.). Deduction, he insisted, must proceed from what is *a priori*, or self-evident. 'As long as we refrain at each stage, from accepting anything as true which is not, and always keep to the order required for deducing one thing from another,' Descartes declared, 'there can be nothing too remote to be reached in the end or too well-hidden to be discovered'.*

That was all very well as long as the first principle in the chain of reason – i.e. the one doing the reasoning – was beyond doubt. The evidence of one's senses was not, of itself, proof. After all, the movements of the heavenly bodies had appeared perfectly obvious before the telescope came along and proved perception to be wrong. This led Descartes to the proposition everyone knows about him: '*Cogito ergo sum*', 'I think, therefore I am'. If I doubt everything, I cannot doubt the fact that I am doubting. Therefore, I am not a dream. Life is not a mirage. But what about the ultimate Reality? Could Descartes, with equal certainty, prove the existence of God? Here his thinking was, perhaps unconsciously, following that of earlier Christian apologists such as Augustine and Aquinas, though he worked it through more intricately.

He offered three proofs:

* J. Cottingham, R. Stoothoff & D. Murdoch (eds.), *The Philosophical Writings of Descartes*, CUP, Cambridge, 1984–1985, II, 1:120

1. The idea of God as infinite and perfect cannot derive from man, since man is finite and imperfect. Since every effect must have a cause, the idea of God must derive from God.
2. A finite and imperfect being capable of conceiving of God cannot be brought into being by anything that is also finite and imperfect.
3. If God did not possess the attributes of infinity and perfection ascribed to him he would not exist.

Was Descartes' God the Christian God or a remote idea – even a personal creation of his own philosophical method? His starting point was not revelation as presented by the Bible or the doctors of the Church. As we have seen, he had an inbuilt mistrust of ecclesiastical authority. Deists, who believed in an original 'cause' who created the universe and then left it to run like a clockwork automaton, would later claim him as one of their own and critics charged that he had reduced God to a 'mathematical abstraction'. Such objections do not take fully into account Descartes' historical milieu. The only supreme being he was unable to doubt was the perfect and infinite God, who, as well as being responsible for the infinitely complex operation of the material universe, was also the source of all ethics and wisdom, of whom both Catholics and Protestants spoke (and in whose name they were currently slaughtering each other on European battlefields). He did not, like some earlier seekers, embark on a quest for a hermetic deity pre-existing or embracing the recognised world religions. It was essential to his system that the God of whom he was aware was *knowable* and *involved* in the affairs of his human creatures.

This necessiated an understanding of the physical and spiritual make-up of man. Descartes' explanation was a metaphysical dualism in many ways similar to that of Plato. Mind (soul) and body are distinct. Each has its own basic 'notions'. Those of the body concern shape and movement. Those of the soul have to do with consciousness. Like earlier philosophers Descartes grappled with the relation between these two entities. How does the mind control the body's behaviour and how does the body react on the soul to cause sensation and emotion? His predecessors had located the soul in one of the major organs or the blood. Descartes suggested that it resided in the pineal gland of the brain. However, he was not asserting that mind and brain are mutually dependent and unable to exist separately. On the contrary, his principal aim was to show that the soul was not tied to the body. Being a different 'notion' it would not cease to exist after the body's decay. He actually stated in *Meditations on First Philosophy* (1641) that his intention was to convince those who could not be swayed by theological argument that the soul is immortal (and thereby to dispose them to a moral life).

However, there remained an ambiguity about Descartes' Catholicism that pursued him even after death. He was buried in the cemetery of an orphan's hospital and in a part of the graveyard reserved for the interment of children who had died before baptism and whose eternal destiny was problematic. Apparently the Church was as uncertain about Descartes as Descartes was about the Church. Rome's final verdict was given in 1663. The Vatican censors were generally out of sympathy with Descartes' anti-Aristotelian natural philosophy but it was not his Copernicanism that most worried them. At several points his dualism seemed to impinge on theology. For example, it apparently cast doubt on official teaching about the human soul and in his understanding of matter there was no room for the separation of 'accidents' and 'substance' on which the doctrine of the Eucharist rested. Several of the philosopher's publications were placed on the Index.

Another Frenchman with a distinctive way of thinking was Blaise Pascal (1623–1662). Indeed, the classic of Christian apologetics he left the world has always been known as *Pensées – Thoughts*. But there any similarity ends. Pascal was dismissive of the older Frenchman's entire approach. Descartes' god was a cold abstraction; not the God of love Pascal claimed to know:

> The metaphysical proofs for the existence of God are so remote from human reasoning and so involved that they make little impact and, even if they did help some people, it would only be for the moment during which they watched the demonstration, because an hour later they would be afraid they had made a mistake.*

It may be that Pascal recalled an earlier snub he had received from Descartes. At the age of sixteen the boy had actually ventured into print with his solution to a problem in geometry. The philosopher first announced that the treatise had obviously been written by Pascal's father. Then, when he had been disabused of this error, he observed huffily that it was not difficult to improve on the work of the Greeks.

Pascal was a prodigy, the son of a Normandy tax collector and amateur scientist. His father personally attended to the education of his children and obviously made a good job of it, for Blaise and his two sisters were all highly intelligent. The family moved to Paris when Blaise was still a child and, as he grew up, he had the opportunity to meet some of the most stimulating thinkers in the kingdom. His first interests were mathematics and mechanics. Before the age of twenty he had invented a calculating machine. He was one of the first to work on probability theory and he did much of the groundwork for the development of calculus.

In 1648 he attracted considerable attention by contributing to the solution of one of the puzzles to which international

* *Pascal's Pensées*, trs. A .J. Krailsheimer, Penguin, London, 1966, p. 86

scientists were addressing themselves – air pressure. According to two Aristotelian assumptions, air was weightless and 'nature abhors a vacuum'. Galileo and his close associates questioned both propositions. It was observation of a practical problem that had suggested the need for a rethink: why could water not be pumped up from a depth of more than 34 feet? Other Italian scientists were working on the problem but it was Galileo's friend, Evangelista Torricelli, who moved the problem to the laboratory. He filled a glass tube with mercury and held a finger over the opening. Then he upended the tube in a bowl of mercury and removed his finger. Immediately the column dropped to 76 cm. Whenever the experiment was repeated the result was always the same. It would seem that a vacuum had been created in the tube and that the height of the mercury column was determined by the downward pressure (weight) of the outside air. News of these experiments spread across Europe and, in France, both Descartes and Pascal interested themselves in them. Both approached the problem differently.

Descartes, the rationalist, could not conceive of 'nothing'. Every created thing has 'substance'. Even that which we cannot perceive with our senses has material existence and is made up of infinitesimal particles. To this extent he agreed with Aristotle that a vacuum is inconceivable. To make sense of the apparent gap in the mercury tube he suggested that it was caused by emanations from the fluid or through the glass.

Pascal, on the other hand, took the inductive approach and set his mind to devising more precise experiments. He carried out variations of Torricelli's original work, using water, wine and other fluids, as well as mercury and different-shaped containers. One fact that immediately became apparent was that the size of the gap at the top of the tube varied according to the specific density of the fluid. He developed a theory that would explain every aspect of the phenomenon. Around autumn 1647, Descartes paid him a surprise visit. The two men discussed their theories – apparently reasonably amicably and the older man actually prescribed some treatment for Pascal's medical condition.

Now Pascal worked out a new experiment – his most famous. It was one at which he could not be present. Unable to travel because of his indifferent health, he enlisted the aid of his brother-in-law, Florin Périer, who lived at Clermont-Ferrand in France's Massif Central. For what Pascal had in mind a mountain was necessary. He had Florin repeat Torricelli's experiment on top of the impressive Puy de Dôme summit, which stands at 1,465 metres above sea level and about 1,000 metres above Clermont-Ferrand. Périer and his assistants carried out the test in town and carefully noted their findings. When they then did it all over again on the peak they saw a marked difference in the height of the mercury – 84 millimetres. The only explanation was that air pressure, which decreases with altitude, was allowing the mercury to settle at a lower level. To double-check his results, Pascal took his apparatus (which we can describe as a primitive barometer) to one of the Paris churches. Even though the tower was only a hundred metres in height, it was sufficient to show a difference in the mercury column between experiments carried out at the top and those performed at ground level.

Not everyone was convinced by Pascal's deductions and debate continued for decades between rationalists and empiricists; those intent on proving theories and those who devoted themselves to further experimentation. For example, the Jesuit scholar, Étienne Noel, argued that since light passed through the space above the mercury column in the transparent tube there must be 'pores' in the glass allowing other atmospheric particles to infiltrate. Meanwhile, some empiricists explored other properties that seemed to show the existence of a vacuum.

Otto von Guericke (1602–1686), a citizen of Magdeburg, was lucky to live longer than his thirties. He was a rare survivor of the appalling 1631 siege which cost the lives of 25,000 of his fellow citizens. Most of his time thereafter was taken up in the work of reconstruction and civic recovery but, as an enthusiastic inventor, he also beavered away at the vacuum problem and its practical applications. In 1654 he created the vacuum pump. Three years later he demonstrated the force of a vacuum in another of those dramatic experiments that enliven the history of science. He placed two copper hemispheres together and pumped out the air. Now, harnessing two teams of horses, one to each side, he attempted to pull the semi-domes apart. It was impossible.

Now the Englishman Robert Boyle (1627–1691) weighed in with another series of demonstrations. He and his assistant, Robert Hooke (1635–1703), were able to construct a glass sphere strong enough to allow the air to be taken out. By placing a variety of objects within the sphere they were able to show that the atmosphere would not convey sound or permit combustion and that the mercury in Torricelli's tube fell dramatically. In the meticulous notes he kept on every experiment, Boyle showed himself to be a model example of Bacon's methodology and was a founder of modern experimental science. Inevitably, argument rolled on. Boyle became impatient with it. The existence or otherwise of a vacuum, he said, was 'rather a metaphysical than a physiological question, which therefore we shall here no longer debate, finding it very difficult either to satisfy naturalists with this Cartesian notion of a body or to manifest wherein it is erroneous'.*

To return to Pascal, he found academic argument between irreconcilables equally frustrating but for different reasons:

* *The Works of Robert Boyle*, M. Hunter and E. Davis, eds. 1999–2000, Pickering & Chatto, London, I, p. 198

> If we submit everything to reason our religion will be left with nothing mysterious and supernatural. If we offend the principles of reason our religion will be absurd and ridiculous.

By the age of thirty-one Pascal had reached the conclusion that, as all branches of the Church had always insisted, one could only enter an experience of God by responding to his own self-revelation in Jesus Christ, a response he made in 1654. To remind him of his conversion he had a parchment fragment sewn into his clothing with these words on it:

> God of Abraham, God of Isaac, God of Jacob – not of philosophers and scholars.
> Certainty, certainty, heartfelt joy, peace.
> God of Jesus Christ.
> God of Jesus Christ.
> My God and your God.
> Your God shall be my God.
> The world forgotten and everything except God.
> He can only be found by the ways taught in the gospels.
> Let me not be cut off from him forever! 'And this is life eternal, that they may know you, the only true God and Jesus Christ whom you have sent.' Jesus Christ. Jesus Christ.*

Pascal's spiritual breakthrough came as a result of his involvement with the Jansenists, a small Pietist movement within the Catholic Church that was viewed with suspicion by Rome and would eventually be closed down. The Jansenists were centred at the abbey of Port-Royal in Paris and one of Pascal's siblings was a nun in the sister house there. The movement's theological impetus came from

* Krailsheimer, op. cit., p. 309

the writings of St Augustine and, in particular, his teaching on the old problem of God's prevenient grace and human free will. The insistence on predestination, justification by only faith and man's inability to attain salvation unless God enabled him to have true repentance brought Jansenism close to the Calvinist position. The Catholic hierarchy and particularly the Jesuits smelled heresy.

A former abbot of Port-Royal had been imprisoned on the orders of Cardinal Richelieu and only released shortly before his death in 1643. As is often the case, martyrdom strengthened the radical community, whose members displayed an intense fervour and commitment to personal holiness that stirred others either to admiration or hostility. Pascal experienced both extremes. He was first drawn to it in his early twenties but later experienced a reaction and all but rejected his faith. His conversion in November 1654 came with the force and vividness not uncommon among people with a deep conviction of sin. During the remaining seven years and nine months of his life he devoted most of his creative energy to theological and devotional writing.

For Pascal, religion was a matter of the heart rather than the head. He argued that to demonstrate God from the 'book of nature' was a laudable ambition but one doomed to failure. Of those who thought to make science a pathway to religious conviction he had this to say:

> I should not be astonished at their enterprise if they were addressing their argument to the faithful, for it is certain that those who have the living faith in their heart see at once that all existence is none other than the work of the God whom they adore. But for those in whom this light is extinguished . . . who, seeking with all their light whatever they see in nature that can bring them to this knowledge, find only obscurity and darkness. To tell them that they have only to look at the smallest things which surround them and they will see God openly;

> to give them as a complete proof of this great and important matter the course of the moon and the planets and to claim to have concluded the proof with such an argument, is to give them ground for believing that the proofs of our religion are very weak. And I see by reason and experience that nothing is more calculated to arouse their contempt.*

Pascal was among those who rejected the Cartesian method of achieving certainty through doubting everything that cannot be proved. Should the doubter doubt whether he is awake or whether his existence is a dream? he asked. Should he doubt whether he is doubting or whether doubt itself exists? Scepticism must have seemed to him rather like the new craze of the fairground carousel, forever moving but getting nowhere. Worse still, it was a carousel revolving in a fog. Pascal insisted that the intellect has to have some kind of anchor. For him that anchor was faith in a God who had resolved our uncertainties by revealing himself. The only proof lay in experience. It was while working on probability theory – the chances of winning or losing at the gaming tables (he invented what was probably the first roulette wheel in order to assist his calculations) – that he proposed, en passant, what has come to be called 'Pascal's wager'. It confronted the rationalist with a serious flaw in the methodology of doubt. Since every human being has to make up his/her own mind about the existence of God, since certainty is empirically impossible and since the stakes are infinitely high, the only rational course to adopt is to believe that God does exist. To the rejoinder, 'But I'm made in such a way that I cannot believe, yet I'm forced to gamble,' Pascal's advice was, 'Seek out the divine spark within, the intuition that reason has its limits. Act on that and experience will bring passion and reason into alignment.'

* Blaise Pascal, *Thoughts, Letters, Minor Works*, Harvard University Press, Massachusetts, 1910, IV, p. 242

One target for Pascal's pen in his later years was the Jesuits. Their casuistry seemed to him to have much in common with the rationalists' complex arguments: they, too, indulged in convoluted reasoning but in the Jesuits' case their objective was not the discovery of truth but the proof of whatever proposition they chose to defend. Pascal's tirade appeared in *Provincial Letters*, published in 1656–1657. When the work came to the attention of Louis XIV he ordered it to be burned. Fortunately, Pascal had published anonymously. He was by then working on a major apologia of the Christian faith. This was destined to remain unfinished at the time of his death. What survived – and what has proved to be his major theological/devotional legacy – was the *Pensées*, copious notes for the unwritten book. It was this that was destined to be the best-known product of his amazing mind, in which generations of readers, scientists and non-scientists, believers and unbelievers have found wisdom. It reveals to us Pascal the sceptic – of sceptics:

> The heart has its order, the mind has its own, which uses principles and demonstrations. The heart has a different one. We do not prove that we ought to be loved by setting out in order the causes of love; that would be absurd.*

In the year that the first *Provincial Letters* appeared, a young Dutch Jew was expelled from his synagogue in Amsterdam. Baruch Spinoza (1632–1677), like Pascal, was plagued by poor health and he, too, suffered from the opposition of those in authority. That rejection could hardly have been expressed more violently than in his condemnation by the leaders of the Sephardic community:

* *Pascal's Pensées* (1669), trs. W. F. Trotter, 2011, p. 277

> Cursed be he by day and cursed be he by night; cursed be he when he lies down and cursed be he when he rises up; cursed be he when he goes out, and cursed be he when he comes in. The Lord will not spare him; the anger and wrath of the Lord will rage against this man and bring upon him all the curses which are written in this book and the Lord will blot out his name from under heaven and the Lord will separate him to his injury from all the tribes of Israel with all the curses of the covenant which are written in the Book of the Law . . . We order that no-one should communicate with him orally or in writing or show him any favour or stay with him under the same roof or within four ells [approximately 180 cm] of him or read anything composed or written by him.

What can this twenty-three-year-old possibly have done to merit such censure from his own people? The honest answer is that we do not know. We may reasonably conjecture that his synagogue elders were ultra-sensitive at this time to avoid criticism by the Christian government under whose authority they lived. The Sephardic community had arrived in the Netherlands as refugees from Portugal and did not wish to be forced out of their haven by the irresponsible behaviour of wild and unorthodox spirits. Only the previous year a sensational and disruptive best-seller written by another Sephardic Jew (since converted to Protestantism) had shaken the intellectual world.

Prae-Adamitae (*Men Before Adam*), by Isaac La Peyrère, was an apocalyptic call to Jews and Christians to unite in adoption of a new theology whose central tenet was that there had existed other races before the creation of Adam. The fact that La Peyrère's book was publicly burned in Paris must have increased its attraction to young Spinoza and his radical friends and it is difficult to believe that he had not read it. Whatever the truth of the situation, the leaders of the Amsterdam Jewish community had much to be nervous about.

This was an era of dissent. The Thirty Years' War had left mainland Europe in chaos. In 1648 there were revolutions in France, Portugal, Naples and Catalonia. Months later, England's Civil War ended with the execution of the king and caused political and religious rifts that would take generations to close. Increasing numbers, despairing of establishing faith communities in the old world, took ship for the new. Inevitably, people – especially young people – were discussing political ideas.

William Wordsworth would later write of another time of heady idealism, the 1789 French Revolution, 'Bliss was it in that dawn to be alive, But to be young was very heaven', and a similar mood was to be felt in the taverns of most cities in the middle years of the seventeenth century. Convictions people had not dared to voice for fear of the stake or the prison cell were now aired more openly. For many thinkers, philosophy began to take on a political aspect, as we shall see in the next chapter.

Spinoza was the son of an Amsterdam merchant and worked with his father in the business. Denied a university education because of his religion, he appears to have been largely self-taught. He gave early evidence of a brilliant mind by mastering several languages. He was, however, instructed in Latin by a strange freethinker by the name of Franciscus van den Enden, who had been thrown out of the Jesuit order. He ran an art shop in Amsterdam and he and his pupil were both members of the fellowship of young Dutch radicals, though whether he or Spinoza was the dominant member of the group has never been established. We can, however, chart his later career.

Van den Enden wrote some political tracts expressing ideas well in advance of their time, including democracy and anti-slavery. He drew up plans for a utopian settlement in North America but never put them into practice. In 1671, he moved to Paris where he was, again, a member of a radical clique. In 1674 he was arrested for complicity in a plot against Louis XIV and was hanged.

Spinoza, meanwhile, maintained a low profile, working as a lens polisher and leading, what most acquaintances agreed, was an exemplary life, ascetic and frugal in its self-denial. It was his writings that provoked both admiration and hostility. He was determined to follow the same rigidly rationalist path as Descartes but detected certain flaws in the Cartesian system that he set out to correct. In the *Geometric Exposition of Parts I and II of Descarte's Principles of Philosophy* (1663) he rejected dualism of mind and body. He asserted that there was only one 'substance' in the universe; everything that exists is 'God or Nature' (*Deus sive Natura)* and humanity is part of it. He rejected the Judaeo-Christian concept of a creator who stood outside what man perceives as reality. God *is* reality and reality is an ever-unfolding sequence of cause and effect that human beings cannot change. Their only means of attaining happiness is by understanding and accepting this. The theological and moral implications of this were worked out in his *Theologico-Political Treatise* (1670) and *Ethics* (published posthumously).

The intellectual challenge Spinoza presented, not just to religious orthodoxy, but to disciples of Descartes and to the whole tenor of conventional thinking, raised a storm of protest. The Dutch government banned *Theologico-Political Treatise* in 1674 and it appeared on the Catholic Index not long afterwards – it is not surprising that Spinoza had published the book anonymously. Pursuing his own line of reasoning had led him into a wilderness where few could or would follow. Among his most challenging conclusions were:

1 Nature is deterministic and unchangeable; everything proceeds by cause and effect.
2 Man has no free will and no ability by prayer or any other means to change events.

3 Everything emanates from God; good and evil are meaningless concepts.
4 There is no revelation apart from Nature; the Bible is just another book.
5 God does not act against his own nature; miracles do not happen.
6 Men cannot acquire knowledge; their minds are contained in the divine mind.

Very few late-seventeenth-century thinkers were prepared to follow Spinoza down his pantheistic path, though his philosophy was used as a seedbed a century or more later by poets of the Romantic period. Some contemporaries – atheists and angry anti-establishment rebels – cherry-picked his *opera* in support of their own manifestos, but they were not prepared to put their heads above the parapet and engage in open debate.

When a deliberately offensive work entitled *A Treatise of the Three Imposters* appeared around 1680 it only circulated in manuscript. It 'exposed' Moses, Jesus and Mohammed and dismissed as nonsense all ideas of God, the devil, the soul, heaven and hell. If such canards point to anything it is the existence of an angry – and perhaps sizeable – underground movement of alienated people who were unable to accept ancient certainties but desperately wanted to create their own new ones. Some were following the rationalists at a distance, often without clear understanding. That is not surprising. Thinkers like Descartes and Spinoza, with their tortuous reasoning, scarcely raised a standard that could be easily followed. They were trying to speak to a generation who were not so much unbelieving, as confused:

> When children of the Reformation and Counter-Reformation and children of the Jewish Diaspora turned on the religions

> which had bred them, they mostly sought not to abolish God but to see him in a clearer light.*

Diarmaid MacCulloch's analysis is convincing. In all European countries the authority of the nation-church had been overthrown or, at least, seriously questioned. The old familiar rituals had disappeared or were losing their psychological grip on many minds. There was little respect for spiritual leaders who used state-backed coercion to maintain themselves in power. The Protestant assault on old truths and the Catholic overreaction had contributed to political upheaval, war and sickening atrocities. Yet atheism was too frightening a prospect for the majority of ordinary folk. The only answer it offered to the age-old question, 'What must I do to be saved?' was 'There is no answer'. How 'ordinary folk' responded to and contributed to popular religion/magic/superstition is something we will address in the next chapter. Now, we have one more philosopher to consider – a man who tried to bridge the gap between rationalism and traditional Christianity.

Gottfried Leibniz (1646–1716) was, of all the seventeenth-century philosophers, the one with the best formal education and the most prominent political position. He was born in Leipzig, the son of a university professor. His father had died in 1653 but the boy had the advantage of inheriting an extensive library to feed his hunger for knowledge. By the age of twelve he had mastered Latin among other languages and, three years later, he entered university. By the age of twenty he had obtained his degree in philosophy and also qualified in law. He eschewed an academic career in favour of entering the civil service. In 1677, by dint of hard work and useful patronage, he had achieved the position of Privy Councillor of Justice to the ducal court of Brunswick. Thereafter, he enjoyed the life of courtier,

* D. MacCulloch, op. cit., p. 698

adviser, diplomat and record-keeper but his indulgent employers allowed him to devote time and energy to travel and private study. He became a member of several learned societies and exchanged ideas through letters and private meetings with most of the age's leading thinkers, including Spinoza, whom he met in 1676.

As a result of having to attend to his professional duties his contributions to the history of thought largely took the form of brief articles and letters. Despite this, many of his ideas were ahead of their time. Bertrand Russell awarded Leibniz the accolade of having developed logic in ways that would not be understood for another two hundred years. But here we are concerned with the man *and his time*. We will attempt to tease out what trends in contemporary thought he was responding to and what he wanted to achieve. This will mean setting aside the contributions of this polymath to several branches of mathematics, not because they are unimportant but because they would require several chapters of closely reasoned analysis and move us away from our central concern, the relationship of science, religion and superstition.

As a philosopher and a Protestant, Leibniz's main interest was to bridge the gap between rationalism and biblical Christianity. He tried to show how belief in a transcendent God could be squared with a universe that can be rationally understood. His theological ideas were set out principally in *Theodicy: Essays on the Goodness of God, the Freedom of Man and the Origin of Evil* (1710) and *Monadology* (1714). He rejected Spinoza's one substance proposition, arguing that there must be a necessary and transcendent first cause (God). If the universe is intelligible (and, if it is not, then all thinking about it is pointless), then God is the source of its intelligibility (the infinite intelligence). From this Leibniz deduced that it must be possible to explore why the universe was created in its present form. If, he proposed, God is perfection – both in the sense of being complete and also morally flawless – then what he has created must be the best that could be created. Therefore, our world is the best of all possible worlds. This leads to the two old and still contended issues: the existence of evil and the possibility of human free will.

Leibniz insists that what man calls 'good' and 'evil' must both emanate from God. It is only because man lacks the knowledge to grasp the reality of these entities that he categorises (for example) natural calamities as evil. And it is only because he lacks the will always to choose what is best that he inflicts moral evil and suffering on the world. God permits this state of affairs so that his creatures will, by trial and error, learn to align themselves with the good. They are, thus, possessed of free will. Although God is fully conversant with everything that has been, that is and that ever will be, he does not predestine the fate of any human. Leibniz avoided the determinism of Spinoza and the predestination of Pascal. However, there is a kind of passivity about his concept of the good life, which consists in man aligning himself with the will of God.

Leibniz's insistence that this is the best of all possible worlds laid him open to ridicule by Enlightenment wits – most famously Voltaire. In 1759, the French cynic published a novella entitled *Candide, Or The Optimist*. It recounts the misadventures of its naïve eponymous hero as he tries to live by the philosophy of his tutor, Dr Pangloss. He experiences a wide range of appalling tragedies, encountering natural disasters and every kind of human depravity. Part of the story is set in the aftermath of the Lisbon earthquake of 1755, a cataclysm of biblical proportions in which anything from ten thousand to one hundred thousand people perished in the quake and the tsunami and fires which followed. Such events fail to shake Pangloss's optimism and Voltaire exposes the tortured reasoning the teacher uses to defend a philosophy that is 'obviously' misguided, pouring scorn on all metaphysical speculation and concluding that all we can do is *cultiver notre jardin* – simply try to make life as pleasant as we can for ourselves and others. This reflection was later echoed with a Christian gloss in the prayer of theologian Reinhold Niehbur: 'Lord, give me the serenity to accept what I cannot change, the courage to change what I can change, and the wisdom to know the difference.'

The Frenchman's criticism of rationalism (he also lampooned other philosophers) has a point, though philosophers would probably respond that thinking men are under moral obligation to use their intellect to probe the meaning of existence. Leibniz would doubtless have defended his position by pointing out that though this world is manifestly not the best – i.e. the perfect, happiest, most enjoyable environment for all its inhabitants – it is superior to any other that might be conceived as the home of a limited species blessed with free will and, therefore, interactive with nature. A perfect world is inconceivable partly because people are very rarely 'disposed to wish for that which God desires'.

In defining 'substance', Leibniz reached back to scholastic and Aristotelian concepts while carrying those concepts further. His attitude to the pro-rationalist philosophical giants was that they were essentially on the right road but had not followed it to its logical end. The word he invented for the basic unit of Creation was 'monad'. This lay under the physical 'stuff' of which everything in the universe was composed and also determined the 'form' of everything in it – that which gave it its identity. Monads (apart from God) existed in three ascending categories: all inanimate things; 'souls' (e.g. animals) possessing perception and memory; and spirits or rational souls (including humankind). The monad, a metaphysical term, should not be confused with the atom, which Leibniz's contemporaries perceived to be the smallest component of all matter. Substance, which, perhaps we might call 'is-ness', is the ultimate reality composed of monads, which are 'beings of psychic activity endowed with perception'. They alone have absolute reality. Material objects are merely perceived phenomena.

Like many pioneer thinkers before him, Leibniz understood mathematics to provide the framework for all philosophical exploration. He made a remarkable number of breakthroughs in various branches. He is probably most remembered for his long and acrimonious argument with Isaac Newton over which of them developed calculus (see p. 254) but he was also responsible for advances in algebra, geometry, the binary system and topology. He perfected a calculating machine that was more complex and multi-functional than Pascal's. Leibniz was able to apply his mind to problems in almost any field of intellectual endeavour. The astonishing breadth of his expertise and inventiveness marked him out as one of the greatest geniuses – if not the greatest genius – of the age.

Mathematics – to which his mind tended to turn whatever the problem – was an abstract study that had no direct connection with morality or religion. The subject was, even in Leibniz's time, still widely regarded with suspicion. It was thought to be an uncanny, arcane science that expressed itself in numbers, codes and symbols and looked much too much like magic. In 1651, when John Rowley calculated the height of a church steeple with the aid of geometry the incumbent accused him of conjuring. This was the background against which practitioners operated as they sought to establish the laws governing a wide range of phenomena. This science, as impersonal as it was precise, was seen by many as challenging, rather than glorifying the Creator; not just explaining him, but explaining him away.

Nothing was further from the mind of Leibniz. Rationalism could certainly lead to atheism and for many thinkers of this era it did so. Many deists rejected or were not interested in revelation but only in the God who could be perceived in nature. If we call Leibniz a deist, it was certainly not because he fell into that camp. Like Aquinas, and other Church Fathers, he believed that by reason the truths of Christianity could be made self-evident. Among the many activities of this polymath was an endeavour to reconcile Catholics and Protestants in the Holy Roman Empire. That certainly meant considering the ways the biblical revelation should be interpreted.

Towards the end of his life Leibniz was dismissed by many as yesterday's man. This was partly because he dabbled in so many different aspects of scientific enquiry. Though he had been elected to the philosophical societies of Berlin, Paris and London, he tended to be dismissed by academics as a gadfly or as an intellectual 'politician' more interested in making a name for himself than in engaging in serious debate. His conflict with the much revered Isaac Newton certainly cost him dearly in terms of his reputation among the intelligentsia. It was left to later generations to appreciate his true worth.

CHAPTER NINE

The religion of doctors

The first half of the seventeenth century was a golden age of English prose. Two of the more remarkable texts of the period were unique to the point of eccentricity. Yet what makes them remarkable is that they were extremely popular. They obviously struck a common chord with many erudite readers.

Robert Burton (1577–1640) was a fairly close contemporary of Shakespeare and, like the dramatist, was profoundly interested in the human condition. He was a scholar who spent most of his life as a student (equivalent of a fellow) at Christ Church, Oxford, where he lectured in mathematics and practised astrology. In 1621 he published *The Anatomy of Melancholy: What it is, with all the Kinds, Causes, Symptoms, Prognostics and Several Cures of it. In Three Main Partitions, with their several Sections, Members and Subsections – Philosophically, Medicinally, Historically Opened and Cut Up*. He reissued and enlarged his magnum opus four times and it went on being published long after his death, running to seven reprints in its first fifty years.

Burton's *Melancholy* defies definition. It was a discursive and rambling text displaying a wide knowledge of literature of many kinds – ancient and modern philosophy, theology, medicine, history and poetry. Burton tells the reader that his motivation for writing was that he, himself, suffered from melancholy and sought by discoursing upon it to cure himself. The book is by turns solemnly erudite and self-deprecatingly facetious. From what we can deduce through descriptions of Burton by contemporaries, this matched his own personality: sometimes sombre, sometimes jocular. It may be that he suffered from what we would now call bipolar disorder.

Whether or not that is so, Burton offers a remarkable essay in self-diagnosis. Here was a man suffering from bouts of depression and trying to understand this affliction. The fact that most people, then as now, experience black moods which have no obvious cause must largely explain the long-lasting popularity of this book (among Burton's 'disciples' were John Milton, Samuel Johnson and Charles Lamb). They could identify with the author.

Burton catalogued a range of causes for 'melancholia', including love, hypochondria, reason, superstition and religious frenzy. He first issued a warning to the potential reader:

> . . . that he read not the symptoms or prognostics in the following tract, lest by applying that which he reads to himself, aggravating, appropriating things generally spoken to his own person (as melancholy men for the most part do), he trouble or hurt himself, and get in conclusion more harm than good.*

Then he meandered through the works of the philosophers in search of definitions and explanations. Here, for example, he tackles an old problem, long debated by philosophers and theologians:

> This question of the immortality of the soul is diversely and wonderfully impugned and disputed, especially among the Italians of late . . . The Popes themselves have doubted of it: Leo Decimus [X], that Epicurean pope, as some record of him, caused this question to be discussed *pro* and *con* before him and concluded, at last, as a profane and atheistical moderation, with that verse of Cornelius Gallus . . . 'It began with nothing and with nothing it ends . . . '.†

* *www.exclassics.com/anatomy/anatomy1.pdf*, p. 48
† Ibid., p. 146

Burton, however, devoted little space to theological debate; not, as he explained, because he disdained it. 'I do acknowledge it,' he affirmed, 'to be the queen of professions and to which all the rest are as handmaids'. He simply saw no need: 'There be so many books in that kind, so many commentators, treatises, pamphlets, expositions, sermons that whole teams of oxen cannot draw them.'*

The spirituality that pervades *The Anatomy of Melancholy* is evident throughout. We may think of it as translating into literary form the asclepieia, the temples of healing, popular throughout the Greco-Roman world for more than a millennium (see p. 93), where healing was sought, not only through bodily well-being, but also through psychic harmony. Burton's understanding of illness as resulting from an imbalance of the humours and healing as essentially spiritual was firmly rooted in the teaching of Aristotle and Church tradition.

Meanwhile, medical thinking about the relationship of body and soul (or mind) was developing along new lines in Burton's day. He must have been aware of the work of his contemporary, Rudolph Gökel (Goclenius) of Marburg who, in 1590, gave the world a new word – *psychology*. Of course, the term did not describe a new branch of medicine or even of science. The workings of soul/spirit/mind were still very much the subject of philosophical enquiry. Burton's contribution to the development of embryonic clinical psychoanalysis lay not in advancing new theories about the interconnection of body and intelligence but in anchoring speculation in human experience.

Much the same can be said for that other seventeenth-century literary phenomenon, Sir Thomas Browne (1605–1682), who was writing his most famous book *Religio Medici* while Burton was

* Ibid., p. 46

still revising *The Anatomy of Melancholy*. Like the older man's work, Browne's was essentially autobiographical. Rather than steep himself in ancient philosophy or cross swords with theologians, he looked within himself to discern the truths common to humanity.

Browne was a physician who had studied at Oxford, Montpellier, Padua and Leyden and who then spent forty-six years as a doctor in Norwich. He wrote *Religio Medici: A Physician's Religion* soon after settling back in his own country, around the age of thirty. Probably he wanted to get down on paper his thoughts on 'life, the universe and everything' while they were still fresh in his mind. Certainly, he did not intend the work for publication but simply circulated it in manuscript among his friends. However, a copy found its way, via the Earl of Essex (later a Parliament army leader in the Civil War), into the hands of the courtier and natural philosopher, Sir Kenelm Digby, in 1635. Digby was at the time a most troubled man. Plagued with religious doubts and also under suspicion of having murdered his wife, he had become a virtual hermit, shutting himself away from the world to carry out chemical experiments. It was perhaps to divert him or to assist his reflections on faith and morals that his friend sent Burton's manuscript to him. Digby responded with a detailed critique and it was this which drew attention to *Religio Medici* and led to its publication.

Digby took polite exception to a mere physician's presumption in venturing into the sacred groves of the philosophers and proffering his opinions on such subjects as the immortality of the soul. Yet it was precisely Browne's down-to-earth, layman's observations that won his book the admiration of contemporaries throughout Europe (despite its title, *Religio Medici* was written in English but within months it was translated into Latin and published in London). Here was a practical man of science writing honestly about his religious convictions – and reservations. A loyal Protestant, Browne found the account of Creation in Genesis 'obscure', nor could he conceive how a loving God could tolerate the existence of hell. But he was quite clear on the limits of reason in matters religious: 'To believe only possibilities is not faith, but mere philosophy.'

Like most of his contemporaries, however, he accepted the existence of witchcraft and he is recorded as having given evidence against two women accused of sorcery. But he did reject astrology: 'Burden not the back of Aries, Leo or Taurus with thy faults, nor make Saturn, Mars and Venus guilty of thy follies.' In one of his more revealing statements he wrote, 'Every man hath a double horoscope, one of his humanity, his birth; another of his Christianity, his baptism and from this [i.e. the latter] do I compute or calculate my nativity – not . . . esteeming myself anything before I was my Saviour's, and enrolled in the register of Christ'.

It is salutary to keep in mind that, although the leading philosophers of the age did have some impact on the way physicians (among others) plied their craft, practical medicine and philosophical speculation were moving along paths that were basically diverging and men of faith could be pilgrims on either. Men of medicine did not need the urging of Baconian empiricism or Cartesian rejection of all that cannot be proved to make fresh discoveries. They were involved in the nitty-gritty of bodies, diseases and ailments and they (or at least the best practitioners among them) were open to new understanding of their patients' needs.

Philosophers speculating on the function of the human body tended to fall into two camps: iatrophysicists and iatrochemists. The first thought of the body as a 'machine', whose parts were designed to function together. Disorders could, therefore, be rectified by exploring which units were not working properly. This approach led to the invention of numerous surgical implements to aid investigation and treatment. For example, Santorio Santorio of Padua (1561–1636), made the first clinical thermometer. Giovanni Borelli of Pisa (1608–1679) carried out exhaustive work on the function of muscles in humans and animals, establishing en passant that the heart was a muscle and worked in the same way as others.

Iatrochemists thought of the body as a piece of apparatus in which chemical reactions were consantly taking place. Treatment of malfunctions could, thus, be addressed by medicinal means. Francis de la Boë (1614–1672), otherwise known as Franciscus Sylvius, professor at Leiden, set up the first academic chemical laboratory and carried out extensive exploration of bodily fluids and gases in his search for the mysteries of life and the ways in which alchemy could both unlock those mysteries and regulate the function of the organs. In fact, he may be said to have freed alchemy from the alchemists and the taint of magic and charlatanism.

His close contemporary, Thomas Willis, Professor of Natural History at Oxford (1621–1675), was a staunch royalist and Anglican and, for some years, a physician to Charles I. He is credited with proposing the possibility of intravenous medication but is more remembered for his pioneering work in neurosurgery. A major problem experienced by anatomists was the rapid deterioration of the brain after death. Working with his colleague, Christopher Wren, Willis developed a technique of injecting fluid into the specimens he wished to examine under the microscope to retard deterioration.

What inspired Willis, like many of his followers, was the quest for the link between body and soul. For this reason, he probed the differences between human brains and those of various animals. His work was, of necessity, a combination of observation and speculation. And though he found no answers to the big question and was only able to offer theories concerning depression (melancholia) and other issues engaging the minds of proto-psychologists, he did begin a new chapter in the understanding of the physical structure and activity of the brain.

Any account of seventeenth-century medicine must begin with William Harvey (1578–1657), whose work on the circulation of the blood it is not an exaggeration to call revolutionary. He studied at the University of Padua, Europe's leading medical school, where anatomy lectures were given by Hieronymus Fabricius, who had been a pupil of the great Vesalius. In 1602 Harvey became a fellow of the Royal Society of Physicians and took up a post at St Bartholomew's Hospital in London. He also delivered over several years a course of public anatomy demonstrations and lectures. His detailed notes reveal that, as well as the writings of medical authors, he was familiar with Aristotle, several of the Latin authors, St Augustine and the Bible. His own researches involved dissection of many different animals.

It was very early in his career that he made his groundbreaking discovery. 'The movement of the blood,' he wrote, 'is constantly in a circle and brought about by the beat of the heart.' This was a remarkable assertion and was certain to create controversy. Current understanding was that the blood moved to and fro between the valves of the heart and that the veins carried it around the body. To forestall opposition, Harvey waited twelve years before publishing his findings and spent those years carrying out exhaustive experiments.

This may sound very sensible and obvious but unravelling the workings of the human body engaged anatomists in all manner of experiments that strike us now as bizarre. For example, in the pages of Samuel Pepys' diary there is a report of a blood transfusion carried out between two dogs. On another occasion the diarist recorded how an 'addle-brained man' had received blood transfusions of sheep's blood – and apparently lived, though the experience did not improve his mental capacity. Harvey's *Exercitatio Anatomica De Motu Cordis et Sanguinis in Animalibus* (*Anatomical Exercise on the Motion of the Heart and Blood in Animals*) was published in Frankfurt in 1628. At seventy-two pages in length it was a surprisingly short book for such a momentous subject. The expected protests were not slow in coming, particularly from continental physicians, but Harvey's thoroughness and foresight paid off and within a few years the importance of his discovery was generally acknowledged.

Honours followed recognition but Harvey's career was, inevitably, shaped by the Civil War. Charles I had appointed him to his court as physician-in-ordinary and Harvey moved with the royal retinue to Oxford when it was forced to leave London. As far as anyone could in those days, he avoided controversy. When asked to assist in the trial of four suspected witches, he employed ten midwives to assist in making a thorough search of the defendants' bodies for the telltale 'devil's marks'. He reported that no such blemishes could be found and, as a result, the women were pardoned (though not acquitted). Harvey was present at the Battle of Edgehill (1642) and, although his activities are not reliably recorded, we may assume that he assisted in the care of the wounded. Never greatly interested in politics, he was happy to return to his work at the end of the hostilities.

Harvey's curiosity was boundless. His study of comparative anatomy embraced numerous species from deer (he had licence to dissect animals in the royal parks) to chickens' eggs. His minute observations on the development of the embryo bird covered several years. Once, when accompanying the Earl of Arundel to Germany, the party was delayed because Harvey was nowhere to be found. At last he was tracked down 'making observations of strange trees and plants'. He continued recording and publishing his findings until the end of what was a long and fruitful life. Great though he was in the annals of medical history (some would say the greatest), he knew that he was only a part of the unfolding saga of scientific discovery. For example, he could advance no definite explanations of *why* blood circulates, nor could he put in place the final pieces of experimental evidence to prove that it does. This latter was left to Marcello Malpighi of Bologna to establish. With the aid of the newly perfected microscope he examined a frog's lung and discovered the infinitesimal blood vessels connecting veins and arteries.

Just as the telescope transformed the study of the universe, so the microscope revolutionised the exploration of animal and vegetable species. The principle of optical magnification had been understood for centuries but not until the mid-seventeenth century did scholars possess instruments powerful enough to probe the deepest secrets of human anatomy.

One of the first to make significant discoveries was the Jesuit Professor of Physiology at Würzburg, Athanasius Kircher (1602–1680). He used a microscope to examine the blood of plague victims, publishing his findings in *Scrutinium Pestis* (1658). He identified the presence of minute organisms (probably blood cells) that he called 'animalcules', labelling them the cause of the plague. Though his diagnosis was incorrect, he had established the important principle that contagious disease is caused by minute organisms.

Kircher was yet another polymath or perhaps we should call him a man of incurable curiosity. His interest in geology led to him being lowered into the crater of Vesuvius (then close to erupting) and to reflections on the lost island of Atlantis. He perfected the 'catotrophic lamp' of Christiaan Huygens, which a later age would know as the magic lantern. The sight of images projected onto a screen was a cause of amazement and even fear to beholders and Kircher had to assure them that the process had nothing to do with magic. He was also responsible for musical curiosities such as a water-powered organ. His writings on numerous subjects and his tireless correspondence with other scholars earned him a wide reputation, though towards the end of his life he was dismissed by other scholars because of his wide interests and wild speculations. Some of his initiatives showed him to be a man ahead of his time. For example, he theorised that animals adapted themselves to changing environments. Thus, he suggested that deer migrating northwards became reindeer. Two hundred years before Darwin, Kircher was speculating about the possibility of evolution.

Returning to microscopy obliges us to acknowledge three Dutchmen: Antonie van Leeuwenhoek (1637–1723), Jan Swammerdam (1637–1680) and Steven Blankaart (1650–1704). The new instrument fascinated part-time scientists. It would be unduly cynical to call it a rich man's toy but it did enable those with the time and leisure to indulge their hobby of collecting specimens and subject them to close scrutiny. Leeuwenhoek was one such. He was a wealthy draper of Delft, a busy little centre of trade and culture, famous for Vermeer, tapestry, tin-glazed earthenware and the Delft Thunderclap – a gunpowder store explosion that in 1654 destroyed much of the city. He has been called the father of microbiology, though he never regarded himself as a great pioneer.

Leeuwenhoek comes across to us over the centuries as one of the more appealing characters in the history of science – ardent in his enthusiasm, meticulous in his methodology, clear thinking in deductions, convinced of their accuracy, yet humble in his own self-evaluation. He never wrote a scientific paper and it was left to others to translate his Dutch reports into Latin for wider circulation. He was genuinely astounded, in 1680, to be elected to membership of the Royal Society of London and, although proud of the honour, he never crossed the North Sea to attend meetings.

He was a convinced Calvinist whose ambition was not the pursuit of fame but the demonstration of the intricate wonders of divine Creation. It was only due to the persistence of a friend that he drew the attention of the Royal Society to his early work but, having received their encouragement, he wrote some 560 letters to them during the course of his long life.

The first fact that impressed the English scholars was that this Dutch merchant, a specialist in fabrics and high fashion, ground his own lenses and made his own instruments (some two hundred of them), which were among the finest available. What Leeuwenhoek observed, described and drew in exquisite detail were a number of microbiological 'firsts' – red corpuscles (see p. 210), the striations of muscle fibres and the compound eyes of insects. Remarkable as these achievements were, the greatest sensation was caused by his claim to have observed single-celled organisms, such as bacteria. No one at the time believed in the existence of such fundamental physical units. They found no place in the current understanding of natural philosophy. In 1677 the Royal Society arranged for a delegation comprising committee members and leading Dutch Lutheran pastors to go to Leeuwenhoek's study and see for themselves what the earnest merchant claimed he could see. The result vindicated him completely.

Another scholar who visited Leeuwenhoek at the same time was Gottfried Leibniz. He, too, was impressed:

> Nothing better corroborates the incomparable wisdom of God than the structure of the work of nature, particularly the structure which appears when we study them more closely with a microscope . . . A man in Delft has accomplished wonders at it.*

Clearly Leeuwenhoek's experimental work on the physical basics of the natural world influenced the development of Leibniz's theory of monads.

Jan Swammerdam of Amsterdam was a contemporary of Leeuwenhoek whose life was sadly cut short by malaria. His love affair with the microscope led him to devote all his energies to observing and recording the anatomical structure of insects. At least, all the energies a slender purse would permit. Although he came of good family he fell out with his apothecary father who disapproved

* Leibniz, *Reflections on the Common Concept of Justice*, c. 1702

of Jan's wasting time on a mere hobby. The young man had qualified as a physician at Leiden and subsequently toured other centres of learning before returning to Amsterdam to practise as a doctor. Increasingly, his profession took second place to his entomology.

Traditionally, insects were regarded as scarcely worth investigating and several misconceptions existed about these tiny creatures. One was that larvae and adult creatures were different animals. Swammerdam was able to show conclusively that this was not the case. Transferring his attention to higher forms he made other significant discoveries. For example, he showed that muscular activity in frogs was stimulated by messages from the brain and not from the circulatory system. He was actually the first person to detect the existence of red blood cells, though it was left to Leeuwenhoek to complete this research and write it up in detail. Swammerdam paid several visits to Leeuwenhoek and was in much admiration of his optical instruments. However, he took a distinctly superior attitude towards the man himself: 'He reasons in a very barbarical way, having no academic education,' he wrote to a friend.

Swammerdam published his findings in three books between 1669 and 1674. Then he was caught up in a spiritual crisis that might easily have ended his scientific career. The conflict between traditional religion and rationalism was at its height in Holland during the 1670s. We have already mentioned Baruch Spinoza but he was only one ingredient in the intellectual cauldron bubbling in those years. Another, to whom Swammerdam and other scientists were attracted, was Antoinette Bourignon.

Antoinette came originally from Lille but, in her early twenties, she left and began her career as a charismatic mystic. That was in 1638, when Europe was in political and religious turmoil. There were many like her trying to find peace, security, certainty and truth in a world where such desirables were hard to come by. Exactly what she believed is impossible to know, save that it was one of those radical theologies strong on direct divine inspiration, renunciation of 'worldliness' and belief in the imminent Second Coming. She attracted the attention of many people, including leading scholars, but never succeeded in gathering around her a permanent community of disciples. Swammerdam was just one prominent thinker who for a time fell under her spell.

As a conventional Calvinist, the young scientist, like most of his studious contemporaries, was driven by a religious impulse – a desire to explore the wonders of Creation and give glory to God. It was clear to him that this was the prime responsibility of a man of learning. But, either because of some personal crisis or because of the religious ferment of the times, he found himself looking for a deeper certainty than could be provided by natural philosophy. By 1675 everyone was talking about Antoinette Bourignon, that self-styled prophetess and ascetic, preaching a message of self-denial and hounded from town to town by the authorities for her heresies. Swammerdam visited her little community and was won over by her insistence that salvation lay in abandoning worship of the Creation and devoting oneself to the Creator. Swammerdam recorded his response in the form of a prayer:

> I no longer want to look for you [God] in the woods, the groves, or in the mountains, in the fields, valleys or among the heath or in the rivers, waters or the seas. I wish no more to seek you for turning nature upside down, nor by searching in the incomprehensible wonders of the embroidered entrails of animals, both great and small. For their inscrutable origin, oh my God, is in you alone.*

* Cf. W. van Bunge, ed., *The Early Enlightenment in the Dutch Republic, 1650–1750*, selected papers of a conference held at the Herzog August Bibliothek, Wolfenbüttel 22–23 March 2001, 2003, p. 103

Swammerdam burned some of his papers and followed the prophetess to Denmark. But not for long. Bourignon was soon on the move again. Accused of witchcraft, she fled to Hamburg. Only with her death in 1680 did her ceaseless wanderings come to an end.

Swammerdam returned to Amsterdam and seems to have found the peace he was looking for. He resumed his studies but also expressed his thoughts in religious poetry and prose. His answer to the apparent conflict of faith and reason was similar to that enunciated by Pascal (and, of course, embraced by many men of science who were also Christians): religious truth was apprehended by love. It could be informed and supported by reason but no man could think his way to faith. As he wrote: 'One loses reason forever, if one tries to understand God through reason.'* At the time of his early death he was still gathering notes for his most important work, the *Bible of Nature.* Published in 1737, it has been called the greatest collection of microscopical observations ever made by a single worker.

Steven Blankaart (1650-1704) was another physician of Amsterdam who worked in the same areas as Swammerdam and he, too, was briefly caught up in the Bourignon frenzy. He was an entomologist who took up Willis's technique of injecting fluids into specimens in order to examine blood vesselss. He was particularly noted for observing the capillaries, the most minute of all the cells in the circulatory system. Among his wide-ranging interests were the study of childhood diseases and of a healthy, balanced diet. On this latter subject he did not hesitate to link self-control over food and drink with moral rectitude, observing that people strong-willed enough to practise abstinence made the better citizens. He published on several subjects but his most comprehensive contribution to medical practice was *A Physical Dictionary in Which All the Terms*

* Ibid., p. 102

Relating to Anatomy, Chirurgery, Pharmacy and Chemistry are very Accurately Explained (1679). It was in this textbook that the word 'psychology' was first defined as the study of the soul.

The medical men mentioned in the above pages are only a few members of an army of practitioners who took their avocation to new heights during the seventeenth century. Having broken out of the straightjacket in which Galen and his followers had encased their profession, they embraced that empirical discipline necessary for the successful treatment of patients. We conclude with the career of the doctor who, perhaps, more than any other embodied the no-nonsense attitude towards philosophical speculation.

> I have been very careful to write nothing but what was the product of faithful observation, and neither suffered myself to be deceived by idle speculation, nor have deceived others by obtruding anything upon them but downright matter of fact.*

Such was the claim of Thomas Sydenham (1624–1689), devout Puritan, brave combatant in the Civil War, cantankerous controversialist and unorthodox practitioner. He attracted violent criticism from some leading thinkers of the age, but was friends with such luminaries as John Locke, Robert Boyle and Hans Sloane and was idolised by succeeding generations of medical students.

Sydenham came of a Dorset gentry family who, in the mid-century crisis, were committed to the Parliamentary cause. His own education was interrupted by the war. Although the university records are far from complete it seems that Sydenham began studies at Oxford in 1642 and again in 1647. In the intervening years he served as a cavalry captain, saw military service once more when the conflict flared up again in 1651 and was seriously wounded at least

* Letter, 10 December 1687, British Museum, Add.ME.4376, f.75

once. Thereafter, he took himself to Montpellier to study under the popular Huguenot physician, Charles Barbeyrac. Not until 1663 was he licensed by the Royal College of Physicians and by then he was already practising in London.

His unconventional training ensured that he did not learn his medicine from books and he developed a disdain for academic theory. When asked by a trainee physician to recommend a good textbook, his instant reply was '*Don Quixote*'. Sydenham became famous for such impatient quips. 'Anatomy? Botany? Nonsense!' he retorted on another occasion. 'Go to the bedside; there alone can you learn disease.' When his devoted admirer, Hans Sloane, informed Sydenham that he was about to go to the West Indies to study exotic plants, he was dismissed with the stark advice, 'You would do better to drown yourself in Rosamond's Pond [a favourite location for suicide in London's St James's Park] as you go home.' It goes without saying that he was not interested in iatrophysical or iatrochemical research. He was equally scornful of microscopy, believing that God had equipped the physician with all the senses necessary for diagnosis and treatment.

All this suggests the attitude of a dyed-in-the-wool reactionary and some academics dismissed Sydenham as exactly that. In fact, he was a pioneer of another kind of medicine that looked to the patient for evidence rather than to ancient texts or new theories. Through minute observation he was able to describe symptoms in greater detail than his predecessors. For example, he was the first to distinguish between measles and scarlet fever. He was sparing in his prescription of drugs, preferring to recommend simple solutions such as keeping fever-sufferers cool. He used liquid opium for pain relief and was an advocate of exercise and balanced diet. He was sparing in the use of bleeding (often resorted to by physicians as a cure-all), perhaps because on one occasion in the wars he had been wounded and lost a prodigious amount of blood. By the same token, he wrote the most detailed description of gout, a complaint that afflicted him for many years. Sydenham was noted for his generosity in treating patients irrespective of their wealth or social standing.

A similar spirit motivated his publications, as he explained:

> I have always thought that to have published for the benefit of afflicted mortals any certain method of subduing even the slightest disease, was a matter of greater felicity than the riches of a Tantalus or a Croesus.*

Sydenham wrote five books explaining his discoveries and methods. At first they were taken up more enthusiastically on the Continent but were eventually recognised in his home country as essential reading for students and practitioners, especially *Observationes Medicae* (1676). Even so, as Sydenham's biographer observes,

> His reputation does not depend so much upon the many sensible and effective remedies he helped to introduce as upon the general clinical principles which guided his own practice . . . It was by such means that Thomas Sydenham, rebel, soldier and physician, laid the foundation of clinical medicine.†

* R. G. Latham, ed., *Epistolae responsoriae*, 1848, II, p. 5

† K. Dewhurst, *Dr Thomas Sydenham (1624–1689) His Life and Original Writings*, The Wellcome Historical Medical Library, London, 1966, p. 59

CHAPTER TEN

Spreading the word

Philip Stanhope, Earl of Chesterfield, politician and wit, had little sympathy with intellectuals. In his famous *Letters to His Son*, he not infrequently made this clear.

The 'Dear Boy' (Chesterfield's illegitimate son, abroad on various diplomatic assignments in the middle years of the eighteenth century), was bombarded with advice by his father on how to get on in polite society. Chesterfield was scornful of young gentlemen who went on the Grand Tour, collecting statuary, fossils, rare plants, minute creatures for microscopic examination and anything else that took their fancy:

> The trifling and frivolous mind is always busied, but to little purpose; it takes little objects for great ones and throws away upon trifles that time and attention which only important things deserve. Knick-knacks, butterflies, shells, insects, etc., are the objects of their most serious researches . . . such an employment of time is an absolute loss of it.

Chesterfield was dismissive of scholars, too obsessed with 'deep' matters to participate in polite conversation. Such a man, he asserted,

> . . . takes no part in the general conversation; but, on the contrary, breaks into it from time to time with some start of his own, as if he wakes from a dream. This . . . is a sure indication, either of a mind so weak that it is not able to bear above one object at a time; or so affected that it would be supposed to be wholly engrossed by, and directed to, some very great

> and important objects. Sir Isaac Newton, Mr Locke, and (it may be) five or six more since the creation of the world, may have had a right to [such distraction] from that intrinsic thought which the things they were investigating required. But if a young man . . . who has no such avocations to plead will claim and exercise that right of absence in company, his pretended right should, in my mind, be turned into . . . his perpetual exclusion out of company.*

Such worldly wisdom would, of course, have appalled Newton, Locke and the other leading scholars of their generation. The truths they were trying to establish and over which they were contending were, they believed, for the benefit of all. They were concerned to reach all educated people with the fruits of their intellectual labours. 'All educated people', admittedly, only embraced a small section of European society and a section that was almost exclusively male and confined, for the most part, to those who had received at least a smattering of higher education.

This scientific evangelism by the pioneer thinkers of the seventeenth century certainly bore fruit. From about 1660, the Grand Tour became increasingly part of a gentleman's education. The young men who went to gaze on the architectural survivals of the classical world and to attend the lectures of the celebrity scholars in foreign universities brought back with them souvenirs by the shipload with which to grace their mansions and to fill their *Wunderkammer* (wonder rooms or cabinets of curiosities). From some of these in time would grow great European and American museums. It is this educational movement that we will be exploring in this chapter.

* F.K. Root, ed., *Lord Chesterfield's Letters to His Son and others*, Dutton, New York City, 1957, pp. 57–8; 19

The development of universal scientific knowledge in the universities and the households of wealthy patrons followed a pattern of inevitable evolution. Printed books – and particularly Bibles – spurred individuals to desire knowledge. This led to more young men from affluent homes attending university. For those with a thirst for knowledge, learning did not end there. Hence, the appearance of learned societies. Bruno and Galileo visited one such in Venice that had obviously been established for some time. The development of permanent societies began in the shady subculture of late Renaissance Italy, where philosophers, magi and inquisitors moved through the groves of secret knowledge, keeping a wary eye on each other.

One such organisation was established by Giambattista della Porta (1535–1615). He was a Neapolitan gentleman and a lay member of the Jesuit order with an enquiring mind and the intelligence and wealth to indulge it. His encyclopaedic mind embraced cryptology (an important skill for spreading new ideas with the Inquisition looking over his shoulder), botany, alchemy, engineering, optics and astronomy. He wrote several treatises describing his experiments, calculations and theories. In the 1570s he set up a learned society of like-minded scholars, called the Academy of the Secrets of Nature or, more conveniently, *Otiosi* (Men of Leisure).

This small, eclectic group continued for some years to explore and discuss the latest developments in experimental science but, despite their caution, they were accused of dabbling in the occult and were obliged to close down. However, in 1603 some of the members joined a new body set up in Rome by an enthusiastic eighteen-year-old, Federico Cesi (1585–1630) – the Accademia dei Lincei (or Lincean Academy). This was a deliberate (perhaps defiant) move to keep alive the study of natural history. Its strange name came from Giambattista's *Magic Made Natural*, the cover of which bore the engraving of a lynx and the words, 'examining with lynx-like eyes natural objects so that they may be used' (a proposition not a million miles from Bacon's utilitarian empiricism).

The Lincean Academy was encyclopaedic in its approach to knowledge and included in its membership ecclesiastics, soldiers, courtiers and engineers as well as philosophers. Its manifesto encouraged any to join 'who are eager for real knowledge and will give themselves to the study of nature, especially mathematics'. Federico Cesi's sudden death in 1630 brought this promising society to an end but by then similar bodies were springing up as far afield as Venice, Florence, Seville, Madrid and Paris. In all probability, there were many more informal groups elsewhere that have made little impact on history because they did not sponsor publications or leave records of their meetings. Bodies like the English Royal Society and the French Academy of Sciences were destined to have a greater impact but it is as well to acknowledge that the state-sponsored French Academy was, in fact, the successor to just one such less formal organisation.

The Académie Parisienne was the brainchild of the remarkable Marin Mersenne (1588–1648). He came of peasant stock but received an education at the hands of Jesuits and went on to study theology and Hebrew at the Sorbonne in Paris. The Church still provided the only certain route to eminence for men of humble origin. After teaching philosophy and theology at Nevers, Mersenne returned to Paris in 1620 to devote himself to further study. He was passionate about both his faith and scientific advance. His publications ranged from a commentary on the book of Genesis, to denunciations of occultism and treatises on optics and the frequency of oscillation of stretched string. In *The Truth of the Sciences Versus the Sceptics* (1624), he took on the Pyrrhonists, sceptics who rejected *all* certainties and believed that neither religious revelation nor scientific deduction could offer irrefutable knowledge.

In 1635, Mersenne established the Parisian Academy. As well as conducting meetings for the discussion of all aspects of *scientia*, this body set up an international network linking scores of leading scholars, including Galileo and Descartes. Mercenne died in 1648 but his philosophical community lived on and, eighteen years later, became the founders of the French Académie des Sciences.

This activity was paralleled by what was happening across the Channel. In 1628, Samuel Hartlib (*c.* 1600–1662), a Polish-born refugee who fled from the chaos of the Thirty Years' War, and settled in England. He arrived seeking the freedom to study the main branches of knowledge he was interested in. A convinced Baconian, Hartlib was committed to the 'usefulness' of scientific knowledge in areas as diverse as medicine, agriculture and politics. His overriding passion was education and his declared grand ambition was 'to record all human knowledge and make it universally available for the improvement of all mankind'. Like Mersenne, he established and maintained contact with a large number of scholars throughout Europe. His focus was always on the *practical*. For example, he examined Dutch farming methods closely to encourage scientific crop rotation and the development of more efficient agricultural implements. Hartlib's importance lay in the enthusiasm he brought to his plans for the improvement of society. His interests and ideals had remarkable width but sometimes lacked depth: he accepted many of the ideas of Paracelsus and was a believer in sympathetic medicine.

Like many contemporary idealists, Hartlib was concerned about healing the divisions in society. He corresponded with men of different shades of religious and political opinion in his search for formulae that would provide the basis for a more tolerant regime. It was in this spirit that he sought the aid of John Milton and the Czech theologian and philosopher, Jan Comenius.

Comenius was a pastor of the Protestant group known as the Moravian Brethren (see p. 277) who was forced to flee from persecution. His subsequent travels brought him on a couple of occasions to

England. He and Milton shared the belief that only a system of government-sponsored schools for all children would bring society out of chaos and help to establish the godly commonwealth. In a treatise, written at the prompting of Hartlib, Milton proposed a reformed system that would 'teach the people faith, not without virtue, temperance, modesty, sobriety, parsimony, justice', and to subordinate selfish interests to 'the public peace, liberty and safety'.*

Comenius' vision was, to all interests and purposes, identical: 'The general aim of the entire education of mankind [is] to restore men to the lost image of God – i.e. to the lost perfection of the free will.'†

Both aspiring reformers were scornful of the existing system of privileged education, which was traditionalist, hidebound and designed to preserve the social status quo. Such idealism could not prevail against political realities. Yet, following the restoration of Charles II in 1660, English higher education remained securely ring-fenced and firmly closed to all non-Anglicans. However, the principles championed by Milton and Comenius had an influence on the curricula of the dissenting academies set up in opposition to the university system.

And yet it was, at Oxford, that in 1649 a group came into existence that, a mere thirteen years later, would develop into the Royal Society of London for Improving Natural Knowledge. At first it was just another of the many 'thinking men's clubs' that were springing up around Europe. John Wilkins, Master of Wadham College (later Bishop of Chester), opened his rooms to a weekly gathering known as an 'experimental philosophical club'. The name clearly showed its Baconian emphasis. Its members were actually rebels, turning away from the speculative approach of the university to the hands-on

* Cf. C. Hill, *Milton and the English Revolution*, Faber, London, 1977, p. 148

† Ibid.

empiricism that would seek proofs and build on those proofs, practices and inventions that would benefit mankind. They even referred to themselves as the 'invisible college'.

What raised the Oxford group above the level achieved by most of the others was the quality of its membership. William Petty, the Professor of Anatomy, began regular dissection in the university but later made his name as a pioneer of the science of economics. Seth Ward, Professor of Astronomy, made important contributions to the understanding of planetary motions and went on to become Bishop of Salisbury. Christopher Wren will, of course, always be remembered as the architect of St Paul's Cathedral and several London buildings erected in the aftermath of the Great Fire of London (1666) but, in the 1650s, he worked with Wilkins to construct an eighty-foot telescope. Thomas Willis (see p. 204) carried out chemical experiments and wrote his important treatise, *De Fermentatione*, in 1656 and, from 1660, was Professor of Natural History. However, the major luminaries of what became the Royal Society were Robert Boyle, Robert Hooke and John Locke, and to them we will return shortly.

Membership of the experimental philosophical club grew but, at the same time, some of the founders left Oxford to pursue their careers. They remained united in their commitment to scientific exploration and, in 1660, transferred their basis of operations to Gresham College in London. Under the terms of his will, the Elizabethan merchant and financier Sir Thomas Gresham had left his house in Bishopsgate to the City of London for the setting up of a college, similar to those at the universities but dedicated to more modern curricula and modes of teaching.

The club was very fortunate in acquiring such an excellent base and one free from the restraints and traditions that still tended to restrict freethinking at Oxford. They were also fortunate in having the strong backing of the new king. Charles II was genuinely enthusiastic about scientific experiments and granted the scholarly body a royal charter in 1662. From the following year it was known as the Royal Society of London for Improving Natural Knowledge. The king's support inevitably meant that the society and, indeed, all things scientific became fashionable. Several members of Carolean elite were enrolled and a list of members reads almost like a Who's Who of the great and good.

In 1665, the Royal Society began quarterly publication of its proceedings in *Philosophical Transactions*. This rapidly became a major spur to Europe-wide scientific activity. By the end of the century similar bodies had been set up in Paris and Berlin as semi-government agencies. That same year, a new member of the 'invisible college' arrived in Oxford.

The Honourable Robert Boyle (1627–1691) was a young Anglo-Irish gentleman. Having been left a generous legacy by his father, the Earl of Cork, he was in a position to follow his two passions – religion and science. Joseph Addison, the essayist, poet and politician described Boyle as 'an honour to his country and a more diligent as well as successful inquirer into the works of nature than any other one nation has ever produced'. The greatest clinical teacher of the eighteenth century, Herman Boerhaave, waxed eloquent in Boyle's praise: 'To him we owe the secrets of fire, air, water, animals, vegetables, fossils; so that from his works may be deduced the whole system of natural knowledge.' Later generations regard Robert Boyle, simply, as the 'father of modern chemistry'. Such adulation fits strangely with a man who stammered, enjoyed indifferent health and poor eyesight, was almost excruciatingly modest (he more than once declined a peerage), had a tendency to gullibility and might not have taken up a scientific career had it not been for the encouragement and help of his sister, the redoubtable Katherine, Viscountess Ranelagh.

After four years at Eton, young Robert was sent on the Grand

Tour to finish his education. It was in Italy that he met the aged Galileo and became interested in the big question being probed by European scientists – whether or not a vacuum can exist in nature. He followed the experiments of Torricelli and Pascal. Back in England and too weak to enlist like other family members in the Royalist cause, Boyle needed a purpose in life. He spent much time with Katherine.

This remarkable and independently minded bluestocking was a major figure in society who boldly espoused the Parliamentary cause, belonged to various scholarly debating societies and mixed with many of the leading intellectuals of the day, including Milton and Hartlib. Under the influence of Baconian scholars and primed by his own religious conscience, Robert conceived it to be his duty to pursue scientific studies for the benefit of mankind. However, the rural life, whether in England or Ireland, did not provide him with the facilities for proper experimentation and it was this that prompted him to move to Oxford in 1655, where he became a member of the invisible college. He would be a member of this group and its successor, the Royal Society, to the end of his days.

In Oxford, as we have seen (p. 183), Boyle developed an air pump with the aid of Robert Hooke and was able to demonstrate the properties of air and the behaviour of various objects when deprived of it. When Aristotelians disputed his findings, Boyle followed up with a further series of experiments that enabled him to demonstrate not only the strength of air pressure (what he called the 'spring' of the air) but that there was a mathematical relationship between the pressure and volume of air (and all gases) – known ever since as Boyle's Law.

What was characteristic of his work was his meticulous note-taking. Boyle refused to be labelled as a follower of any school of thought but he was a prime example of the Baconian thesis that observation should lead to the collection of data that can be tested over and again through experimentation, thus leading to the recognition of the laws of nature. For over thirty years there flowed from his pen a prodigious volume of treatises and reports on a variety of topics, including the expansion and force of frozen water, the development of new medicines, magnetism and electricity, gem stones, and the qualities of blood. But the two issues which most engaged his mind were the composition of matter and the relationship of religion and science.

In *The Sceptical Chymist* (1661), Boyle took on, in dialogue form, both the Aristotelian concept of the four elements (earth, air, fire and water) and Paracelsian theory of three humours (salt, sulphur and mercury). Boyle had an open mind on alchemy (or, perhaps, a desire to believe it). It was this that involved him, in the 1670s, with Georges Pierre des Closets, a French fraudster who conned many people in England before retiring and buying a country estate on the proceeds. His pitch was that he was an agent of a secret society in Antioch who were the guardians of valuable esoteric information.

Boyle confided in the plausible Frenchman, who promised access to the hidden knowledge Boyle craved – for a fee, of course. How can it be that one of the great pioneer scientists was so credulous? Paradoxically, the answer may lie in his own meticulous methodology. When presented with a new theory he explored it thoroughly, carrying out experiments where appropriate. He would not dismiss an idea without testing it, even if it seemed contrary to common sense. Boyle held to the corpuscular theory – that matter was composed of particles smaller than atoms. Such corpuscles of different substances could, conceivably, interact. Hence, transmutation of metals was logically possible.

Boyle was certainly not alone in taking seriously claims that later ages have dismissed as foolish or fraudulent. For example, Robert Plot, professor of Chemistry at Oxford and Secretary of the Royal Society, worked with an alchemist in a quest for the philosopher's stone. There have been many times in the history of science when unbiased research and gullibility have kept close company.

Problems of physics and chemistry did not completely absorb Boyle's attention. Issues of theology and spirituality were more important. This deeply introverted man always kept a close check on his personal morality and the state of his soul. He deliberately chose a life of chastity. When he left Oxford in 1668 to live with his sister, he found membership of her Protestant circle much to his liking and he worked with others on such projects as having the Bible translated into Irish and supporting missionary societies in the Orient. Here again, his generosity and enthusiasm opened him to the wiles of fraudsters. Boyle was one of those scholars who was attracted to the mystic Antoinette Bourignon (see p. 210). He also took seriously the claims of Valentine Greatrakes, an Irish faith healer.

It would be easy to interpret Boyle's constant wrestling with the relationship between religious belief and scientific enquiry as the anxieties of a solitary soul but that would be to misconceive the intellectual and spiritual turmoil of the age in which he lived. In 1646 he wrote to a friend:

> If any man has lost his religion, let him repair to London and I'll warrant him, he shall find it: I had almost said too and if any man has a religion, let him but come hither now, and he shall go near to lose it.*

He lamented that there was 'no less than 200 several opinions in points of religion' and, since he came from a family whose members found themselves on both sides of the politico-religious divide, these

* Cf. J. W. Wojcik, *Robert Boyle and the Limits of Reason*, CUP, Cambridge, 1997, p. 20

issues could never be merely matters for detached intellectual speculation.

One of the reasons Boyle remained aloof from fashionable Restoration society, shunning the time-servers of the Stuart court, was his concern at the drift towards deism and atheism. In his last years he devoted much time and energy towards revising an earlier treatise setting forth a Christian natural philosophy. It was published in 1686 as *A Free Enquiry into the Vulgarly Received Notion of Nature.* The timing is significant. James II had been crowned king in April of the previous year and had immediately been faced by rebellions in Scotland and England. Having crushed these, the new king established a standing army, lifted restrictions on his co-religionists and sacked parliament. The emergence of a new Catholic tyranny outraged Protestant opinion and provoked writers of books and pamphlets to declare their opposition of everything James and his supporters stood for.

Boyle's *Free Enquiry* struck at the scholastic theology and natural philosophy which underlay not only much Catholic thinking but also the atheism becoming fashionable in some quarters. Boyle attacked two fundamental 'heresies' about 'vulgarly received notions': that reason had the potential to solve all problems and that there was no distinction between nature and providence. Although a passionate believer in the progress of scientific enquiry, he believed that there would always be vast areas of knowledge that would remain outside man's grasp. Furthermore, he asserted that God had placed limits on what could be grasped by reason alone, unaided by divine revelation. Robert Boyle died on the last day of 1691. In his will he left funds for the delivery of annual lectures for the defence of Christianity against atheism and other 'errors'.

In this dislocated century many pioneering thinkers were less concerned with speculation about the working of the universe than with analysing the workings of human society. Everyone longed for peace and stability, so it is not surprising that, just as philosophers turned away from Aristotelian answers to problems of mathematics, astronomy and physics, so they also rejected concepts of kingship, state and church and went in search of new societal and political fundamentals. They wanted to know what the divine ordering of society was or, indeed, whether any such ordering existed. Two English thinkers came up with fundamentally different answers to such questions.

One thinking man who was not admitted to the Royal Society was Thomas Hobbes (1588–1679). He was a controversial figure in his own day and has remained so ever since. Historians and biographers have never been able to agree about what Hobbes did – or, more importantly, did not – believe. His mature years covered the period from the arrival of the first Stuart king through the rejection and execution of the second and right up to the successful re-establishment of the third. These events were both the background to his life and the motivation for his philosophy. It is important to keep in mind that Hobbes was already in his fifties before the chaos of the Civil War swept over the lands obliterating all the comfortable certainties that cocooned his life. He had grown up in a country where Crown and Church worked in concert to rule as God's agents, suppressing heretical ideas and resisting the pretensions of parliament. By the time Hobbes published his major work, *Leviathan*, the 'good old days' were gone. Monarchy had been abolished and the nation's religious life was a free-for-all of Christian and pseudo-Christian novelties. In 1651 he was, by the standards of his day, an old man and it was now that he tried to formulate a scientific basis for the establishment (or re-establishment) of civilised society.

Thomas Hobbes did not have the best start in life. He was the son of a failed parson who deserted his family while Thomas was still a child. The boy was brought up by an uncle who, in the fullness of time, sent him to Oxford to complete his education. He was a spirited student. He did all the things students do and study did not always come top of the list of priorities. At university Hobbes met William Cavendish, son and heir to the Earl of Devonshire and made such an impression on the younger man that he was appointed to be his tutor/guardian/companion.

Life in the entourage of a wealthy nobleman's son was very agreeable and when, in 1610, he took Cavendish on the Grand Tour they did it in style, sampling all the delights the Continent had to offer for five years. But there were also intellectual challenges and Hobbes was excited by the new movements in natural philosophy that contrasted with the backward-looking Aristotelianism of Oxford. Back in England, Hobbes spent some time as an amanuensis to Francis Bacon. But he was not won over to the empirical method. He still thought in deductive terms; the philosopher should begin with propositions and test them by argument. He maintained his aristocratic contacts and became something of an intellectual guru to a wide circle of amateur philosopher-scientists, not dissimilar to those who had gathered round Raleigh and Percy in the Tower of London (see p. 127). The leader of this scholarly club was another Cavendish, William, Earl (later Duke) of Newcastle, a poet, scholar, sometime royalist military commander and an expert on horsemanship.

One meeting place of this group was the Cotswold manor house of Great Tew, home of Lucius Cary, Viscount Falkland, where the major concern addressed was how to heal the religious breaches that were creating such chaos throughout Europe. The members were acutely aware of the widespread scepticism to which the scandal of division was giving rise. At Great Tew, Anglicans, Puritans and even Catholics were brought together in the belief that toleration and rational dialogue could achieve, if not theological consensus, at least agreement to go on talking in the best humanist tradition. To what extent Hobbes shared this spirit of goodwill we do not know. What is clear is that during the years of the Civil War which soon followed, he came to the conclusion that religious unity (or uniformity) could only be achieved by strong government.

In 1634 Hobbes toured Europe again, this time in charge of the next-generation Earl of Devonshire. During this trip he met Galileo, spent some time with Marin Mersenne and his group, and entered into what would prove to be a long-lasting dispute with Descartes. Both men held to a mechanistic concept of nature – that the physical universe runs on immutable 'clockwork' principles and everything in it can be explained in terms of the movements and collisions of matter. But, whereas Descartes' system left room for a creator outside the machine, Hobbes conceived of nothing beyond the material cosmos.

As wide as Hobbes' philosophical and scientific interests were, in the years after his return, the collapse of the tidy, hierarchical society of which he had been a part raised politics to the top of his list of priorities. Charles I fell out with parliament, tried to live without it and, eventually, went to war in an attempt to bring it to heel. Hobbes ventured into print in defence of the status quo and absolute monarchy. Having nailed his colours to the mast he returned to the Continent rather than risk involvement in the conflict. His sojourn there lasted eleven years (1640–1652), most of which time was spent in Paris. Initially, he had several friends at the Prince of Wales' court in exile and, for a time, he was a tutor to young Charles. He devoted much of his time to developing his ideas about human nature, society and politics. The result was the book by which he is best known, *Leviathan*. When it was finished in May 1651, he sent it to England for publication. This decision was motivated by self-interest. Hobbes was in his sixties, his health was indifferent, the Royalist cause seemed lost and he was no longer *persona grata* in the prince's entourage. Several of Charles' advisers did not trust him and his anti-Catholicism alienated some of the exiles' French supporters. It was time to test the waters in England.

The government there was in a state of flux, with army and parliament urgently seeking to establish peace and harmony but unable to agree how to achieve them. One move that had been agreed was to welcome all who would swear an oath of allegiance to the republican regime. What may well have been the clincher for Hobbes was the Battle of Worcester. In September 1651 Charles made one last bid to regain the throne. With the aid of Scottish troops he made a final attempt to resolve the situation by military means. The result was dismal failure at the Battle of Worcester. Five months later the English parliament passed the Act of Pardon and Oblivion. Lamenting the 'miserable and sad effects' of 'the late unnatural war', the Lords and Commons resolved 'to make no other use of the many victories the Lord in mercy hath vouchsafed', but rather to ensure that 'all rancour and evil will occasioned by the late differences may be buried in perpetual oblivion'. To this end all offences committed before 3 September 1651 were to be freely pardoned. These events formed the background to Hobbes' return home in the winter of 1651–1652. His enemies did not hesitate to accuse Hobbes of opportunism but it is difficult to see what alternative course of action could have been taken by an aged scholar in need of patronage and wanting to end his life in peace and comfort.

What *is* difficult to understand is what precisely Hobbes was advocating in his book. *Leviathan, or the Matter, Form and Power of a Commonwealth Ecclesiastical and Civil* provoked argument at the time and has been much debated ever since. The treatise essays a scientific social theory and presents its arguments with stark logic but, over the centuries it has left readers with questions: was Hobbes an unrepentant royalist? Did he believe that de facto power was the justification for government or did he propose that legitimate rule depended on the consent of the governed? Did he believe that political relationships rested upon some divine schema or was religion, for him, simply an essential prop for the civil power?

Hobbes took a pessimistic, not to say jaundiced, view of human existence, famously observing that it is 'solitary, poor, nasty, brutish and short'.* There is nothing that ennobles this creature – no soul, no *imago dei*, not even a mind, for Hobbes refuses to concede that body and mind are different entities. His materialism is based on what he conceives to be the 'law of nature' that decrees that everything must serve the principle of self-preservation. Given the circumstances of Hobbes' life and the horrors unleashed on Europe by rival religions, such scepticism is easy to understand. From this materialistic view of humanity, it follows that for any corporate entity – community, commonwealth, state – to succeed it must be dedicated to the material well-being of its members. Every such entity Hobbes calls a 'leviathan'.

Why leviathan? Hobbes takes this metaphor from the Old Testament (particularly Job, Chapter 41), which presents this mammoth sea monster as the biggest, fiercest and most powerful animal in all Creation. It is awesome and untameable. It has a decidedly sinister aspect and the name is even used as a synonym for Satan, the enemy of God (Isaiah 27:1). Hobbes gives this name to the state or commonwealth because it is 'that mortal god to which we owe under the immortal God our peace and defence'.

Hobbes makes very free with the name of God and quotes copiously from the Bible, thus giving the impression that he takes religion seriously but the devil is in the detail and, when we look closely, it becomes apparent that he has little knowledge of or interest in theology. His ruler of the monster state is not a deputy of the King of Kings. Hobbes specifically rejected the divine right theory by which Charles I had claimed to rule. He did, however, believe in monarchy. He was a classical scholar and had written a translation of Thucydides' *History of the Peloponnesian Wars* and having, like

* N. Fuller, *Great Books of the Western World*, XXV, Chicago, 1952, p. 85

Aristotle and Plato, considered alternative polities such as democracy and aristocracy, he insisted that rule by one powerful man was the only guarantee of peace and stability. But Hobbes' ruler is not Plato's philosopher-king, a wise and benevolent despot. In fact, he is closer to Machiavelli's prince, a man whose effectiveness is based on naked power.

And yet Hobbes is eager to demonstrate that the authority of the ruler is based on the consent of the ruled. To do so he uses social contract theory. This, too, had its origins in the writings of the ancient philosophers but emerged afresh in the Reformation debate. If every individual, as Luther asserted, is responsible for his/her faith it follows that he/she has a right to hold and live by that faith *and* that no government can coerce religious belief (e.g. by burning heretics). The Dutch jurist Hugo Grotius (1589–1645) extended this by proposing that there is a natural law established by the Creator to which any valid human law must relate and which guarantees basic human rights. Hobbes agreed but proposed a contract (written or tacit) whereby subjects yielded up their rights into the hands of the sovereign, who guarded them on their behalf.

Did he, perhaps, have Cromwell in mind when he made absolutism the gift of the people to the ruler and did he hope to please the regime by this somewhat convoluted reasoning? Unsurprisingly, he pleased nobody. Non-royalists rejected Hobbes' absolutism. Royalists resented the implication that people who had gifted their rights to the king could take them back again – i.e. they could rebel. The social contract debate that Hobbes started (without, himself, resolving) went on to become one of the core threads of political theory to energise thinkers and activists down to the American War of Independence and beyond.

Was Thomas Hobbes an atheist? Some questions about him are easier to answer than that one: Was he a materialist? Yes. Was he a sceptic? Yes. Was he a rationalist? Yes. Was there any room in his philosophy for metaphysics? No. Did he accept arguments from experience? No. Did he allow the validity of revelation? Only in a very restricted sense.

Our original question would be easier to tackle if Hobbes had engaged seriously in abstract theology but that was never his prime concern. For him it was *religion*, corporate religion, the religion of that state that mattered. It had value insofar as it served the purpose of the sovereign and thus, in Hobbes' view, the purpose of the commonwealth. Writing of miracles Hobbes observed:

> A private man has always the liberty, because thought is free, to believe or not believe in his heart . . . But when it comes to confession of that faith, the private reason must submit to the public; that is to say to God's lieutenant.*

The Counter-Reformation had sought to end the disruption caused by competing religious opinions by invoking the principle, *cuius regio, eius religio* – 'the sovereign's religion is the religion of the state' – and it was this that Hobbes advocated. Religion is the buttress and the tool of the civil power. On the title page of *Leviathan* the sovereign wields a sword with one hand and a bishop's crozier with the other. The message could hardly be clearer.

Thomas Hobbes lived – eccentric and cantankerous to the last – to see the return of Charles II and to receive from his old pupil a pension that helped him to spend his last years in reasonable comfort in the country residences of his old patrons, the Cavendishes. But whatever token might be given by the king, Hobbes remained persona non grata among both the political and scientific elite. Many shared the sense of outrage expressed by Edward Hyde, Earl of

* Ibid., pp. 190–191

Clarendon, in a lengthy refutation he published in 1673 of the presumptuous author who had:

> ... taken upon himself to imitate God and created Man after his own likeness, given him all the passions and affections which he finds in himself ... [and] comes at last to institute such a Commonwealth as never was in nature or ever heard of from the beginning of the world.*

It would be left to a more secular age to re-evaluate him.

John Locke (1632–1704) could scarcely have been more different. He is commonly regarded as a major contributor to modern political philosophy but, as his biographer, Professor Spellman, points out, the organisation of secular society was not his main concern:

> God remained the final superior and supreme legislator in a manner quite alien to modern western culture ... Locke's abiding faith in a purposeful universe directed by a God who punishes and rewards, where all things are created to serve a greater glory, where sinfulness and its opposition have unimaginable consequences, suggests that this spokesman for reason and enlightenment in the temporal sphere had rather larger motives for his life's work.†

And again:

> ... paramount among [his] concerns, eclipsing even his interest in the new science was the intractable mystery of eternal

* E. Hyde, *A Brief View and Survey of the Dangerous and Pernicious Errors to Church and State in Mr Hobbes's Book entitled Leviathan*, 1673, pp. 28–9

† M. Spellman, *John Locke*, Macmillan, New York, 1997, p. 8

> life: the range of prospects beyond the grave, the sinner's role in securing the great reward, the nature of one's dependence upon both human and supernatural agency in translating what was an all-too-brief passage on earth into the unending and felicitous communion with the Creator.*

In other words, Locke was asking the same question as Pascal, Luther, the medieval peasant confronted by the lurid doom painting in his parish church, and millions of others over the preceding centuries: 'What must I do to be saved?'

John Locke came from Puritan stock. He began his education at Westminster School in the year that Charles I was taken prisoner by the Parliamentary army, 1647, and went on to Christ Church, Oxford, the year before Oliver Cromwell was proclaimed Lord Protector, 1653. During these years the university was purged of all royalist influences and when Locke was not studying, his student days were spent in impassioned debate with other young men about the events dividing the nation and the clashing principles underlying those events. However, his study time was devoted to medicine and, having completed his first degree course, he applied himself to this discipline, though he did not take his BM until 1675. By then the turbulent years had passed and left the pensive student dissatisfied with both authoritarian traditionalism and irrational radical religious 'enthusiasm' (then the euphemism for 'fanaticism'). He found himself caught between two ideologies. Instinctively, he was in favour of the order imposed by king and church. Yet he believed that religious belief and its expression in modes of worship was a matter of individual choice. In the aftermath of the Restoration, parliament and the episcopate combined to enforce conformity. There was to be one state church. Catholics and Protestant dissenters were excluded

* Ibid., p. 1

from public office and prohibited from taking university degrees. Two thousand members of the clergy were deprived of their livings for refusing to accept the Anglican prayer book.

Locke was very much at the centre of things. In 1667 he became personal physician to Anthony Ashley Cooper, Earl of Shaftesbury, a politician who had served under Cromwell, yet survived the return of the monarchy. Shaftesbury held various government offices and was a major contestant in the political manoeuvrings of the 1670s and 1680s, when he firmly opposed any move towards absolute monarchy and was outspoken in rejecting Charles II's brother, the Catholic Duke of York, as heir to the throne (a conflict known as the Exclusion Crisis). In 1682, fearing prosecution for treason, Shaftesbury fled the country and died soon afterwards.

Locke's relationship with his patron had gone well beyond attending to his health. He was a close confidant of the earl, with whom he discussed philosophy and politics and, in particular, was employed in drawing up a constitution for the colony of Carolina of which Shaftesbury was a lord proprietor. Inevitably, Locke came under suspicion of being involved in the earl's 'treason' and, the year after the earl's departure, he too thought it prudent to leave the country. He remained abroad, his principal refuge being Amsterdam, throughout the reign of James II and only returned after the Glorious Revolution had, with comparative ease, replaced the last Stuart king with his daughter, Mary, and her Dutch husband, William of Orange. This resulted in the establishment of the Protestant succession and it made England a constitutional monarchy.

John Locke is widely regarded as the father of liberal democracy and there is a temptation to examine and evaluate his philosophy in the light of modern Western political *lexis* and *praxis*. Yet, fundamental though Locke's thought would be to later political development, it does not fit precisely into any modern theoretical mould. Furthermore, looking at Locke with hindsight obscures both the originality of his thought and the difficult processes he went through to arrive at the principles he enunciated.

Locke lived in an age in which his country rode an extraordinary politico-religious roller coaster, a time that posed numerous questions to thinking men. But Locke was not just a thinking man; he was the most brilliant creative political philosopher of his age. During his foreign sojourn he mixed with many leading intellectuals, fellow exiles and continental scholars. The fruits of all his reflections and discussions appeared during his last fifteen years. What he left to posterity was a synthesis of political and religious philosophy that carried weight because of its clarity and because its originator was someone who had walked in the corridors of power.

The foundation of his philosophy was laid in the *Essay Concerning Human Understanding* (1689). 'How can we *know* anything?' is the first question he poses and he does so because understanding is the 'most elevated Faculty of the Soul'. Like other Christian commentators, by 'soul' he means the *imago dei*, present uniquely in humankind and enabling communication between creature and Creator. His answer: 'Experience'. Every human being is born with a mind that is a blank slate. On the slate he/she writes the information conveyed by the senses. The mind is not only the storehouse of this information but the means of understanding, interpreting and evaluating it.

In accumulating knowledge the mind reflects on what can be derived from the body's physical environment, but this is not the sum total of the evidence available. God provides another kind of *scientia* that is presented to the mind as revelation. The basic vehicle for this is the Bible. Locke was a lifelong and passionate student of the Christian scriptures. His Bible, preserved in Oxford's Bodleian museum, bristles with annotations and he wrote paraphrases of St Paul's epistles. In presenting his theological synthesis based on religion and revelation he was sometimes accused of straying from orthodoxy, a criticism he repudiated. 'I shall [immediately] condemn and quit any opinion of mine, as soon as I am shown that it is contrary to any revelation in the Holy Scriptures,' he responded to one critic.* The means by which we embrace revealed truth is faith but faith is not the enemy of reason. Quite the reverse: reason enables us the better to understand revealed truth.

> Reason is natural *revelation,* whereby the eternal Father of light, the Fountain of all knowledge, communicates to mankind that portion of truth which he has laid within the reach of their natural faculties. Revelation is natural *reason* enlarged by a new set of discoveries communicated by God immediately [and] vouches the truth of, by the testimony and proofs it gives, that they come from God.†

Locke goes beyond Bacon's 'two books' theory by extending the role of reason in the interpretation of Scripture but he readily acknowledges that there are revealed truths which lie beyond human comprehension and these must be accepted humbly as *verba Dei.*

What follows from Locke's understanding of God and man and their relationship is a complex series of propositions explored in treatises covering individual belief and behaviour as well as corporate responsibility in church and state. The moral life consists, first, in discerning good and evil but reason alone does not enable us to always choose the good. The dilemma had been clearly stated by St Paul in Romans 7 – 'The good I want to do I don't do; the evil I don't desire is what I end up doing; I'm aware of a law at work in my body that is different from the one my mind approves.' Beyond stressing

* Cf. W. M. Spellman, *John Locke*, 1997, p. 53

† P. H. Nidditch, ed., *An Essay Concerning Human Understanding*, OUP, Oxford, 1975, 4.18.2

the importance of education, Locke failed to come up with a solution to motivation which, as he acknowledged, was beyond the power of rationality to achieve.

One reason for this may have been his distrust of religious certainties, which were often misplaced and which usually led to mischief. He condemned both the Restoration churchmen who tried to force all Englishmen into the Anglican mould but he also opposed 'enthusiasm'. Harking back to the sectarian chaos of the Civil War era, he strongly denounced those who expressed strange opinions based entirely, as they claimed, on inner revelation not tested or being refined by reason. What Locke was, in fact, saying was that toleration had its limits. Thus, for example, while everyone should be free to seek the truth aided by the twin lamps of reason and revelation, he called for atheists to be 'shut out of all sober and civil society'. His reason was pragmatic: anyone who did not believe in God could not swear an oath and could not therefore be trusted to play his/ her allotted role in society.

But what is the political entity we call 'society'? Locke had seen kings deposed and reinstated, the rise and fall of republican government, attempts to establish absolutism and the see-saw instability of parliamentary factions. His philosophy of the human condition could not be complete without a discussion of the composition and control of states. This he undertook in the *Two Treatises of Government*, which he published in 1689 but continued to work on through subsequent editions. Like Hobbes, he was appalled by the barbaric behaviour within and between nations that had convulsed Europe. Like Hobbes, he sought to define a natural law, agreeable to reason that would enable all people to live in peace, security and freedom. Like Hobbes, he favoured a bottom-up theory of the rights and responsibilities of subjects. But there any similarity between the two thinkers ends. In his discussion of religion he set reason and revelation side by side. He upheld the right of the individual to believe and worship according to his conscience. This degree of freedom implied the separation of church and state. Locke's sovereign was no absolute monarch able to command the consciences of his subjects. On the contrary, Locke claimed that the will of the majority should prevail – even to the point of rebellion – if a situation arose in which a ruler broke the contract in pursuit of his own, personal ends. This placed Locke firmly on the side of those who worked to supplant James II and choose for themselves another king, William of Orange, who reigned as William III from 1689 to 1702.

Though much troubled by asthma, the aged philosopher outlived his sovereign. He spent his last years as a paying guest in the Essex home of Sir Francis and Lady Damaris Masham. Here, the intellectual celebrity virtually held court, receiving visits from numerous admirers, including Anthony Ashley-Cooper, the 2nd Earl of Shaftesbury and his great patron, and the leading philosopher of the age, Sir Isaac Newton.

CHAPTER ELEVEN

Day star of the Enlightenment or setting sun of the age of superstition?

Charles Maynard Keynes, who spent several years studying some of the voluminous quantity of private papers left by Sir Isaac Newton, told his students that this phenomenal genius was not 'the first of the age of reason. He was the last of the magicians.'

If a study of the relations between superstition, magic, religion, philosophy and science tells us anything it is that there is no single thread connecting irrational belief to informed understanding of 'life, the universe and everything'. Numerous tangled strands have to be teased out and some stubbornly resist separation. As we draw close to the end of that slice of intellectual history to which this book is dedicated, we find ourselves dealing with a human phenomenon who embodied these complications:

> Diligent, sagacious and faithful in his expositions of nature, antiquity and the Holy Scriptures, he vindicated by his philosophy the majesty of God mighty and good and expressed the simplicity of the gospel in his manners. Mortals rejoice that there has existed such and so great an ornament of the human race.

So runs the eulogy in Latin on the monument to Sir Isaac Newton in Westminster Abbey. Those who knew Newton might have raised their eyebrows at the description of his manners as expressing 'the simplicity of the Gospel', for this reclusive scholar was tetchy, argumentative and sensitive in the extreme. Yet none can doubt his importance in the history of human thought. To this day scientists debate whether he or Einstein merits the title of the greatest member of their profession. Yet only in recent years have we begun to realise that he devoted a great deal of mental energy to the study of alchemy and that he wrote much more about 'the majesty of God' than he ever did about gravity.

Isaac Newton (1642–1727) was born into a wealthy yeoman family of Woolsthorpe, Lincolnshire. His father died before the boy was born and it was not surprising that his mother looked to Isaac to assume responsibility for the estate and live the life of a respectable landowner. Fortunately, his early mentors recognised the boy's potential and he was allowed to enrol at King Edward VI's Grammar School in Grantham and, subsequently, at Trinity College, Cambridge. He arrived in 1661 and remained as student, fellow and, eventually, professor, until 1695.

At the university Newton found a philosophical school open-minded to post-Galileon science. The Cambridge Platonists might best be understood as bridge-builders. Living in an age of violently warring principles and ideals, they believed in the power of reason to restore harmony. They endeavoured to find common ground between the Calvinism of the Puritans and the High Church Laudian theology of Charles I's church leaders. Similarly, they sought to hold together faith and reason. But certain philosophical circles could not be squared. The mechanistic understanding of nature with its in-built determinism had to be rejected because it denied human free will.

The Platonists embraced Cartesian dualism, the conviction that mind/spirit and body are distinct. From this they deduced that humanity can perceive eternal moral and spiritual truths existing in the mind of the Creator. Hobbesian materialism was anathema to them. The Platonists did not so much enunciate *a* philosophy as engage with the new trends in philosophical thinking and make connections with the pioneers of the classical world. This climate in the Cambridge intellectual greenhouse nourished and shaped Newton's thinking.

It was not long before Newton's enquiring mind was delving into a variety of subjects, engaging the attention of Europe's leading natural philosophers. His notebooks reveal interest in optics, the movement of heavenly and terrestrial bodies, the method of conducting mathematical investigation that later came to be called 'calculus', and sundry problems in geometry, alchemy and theology. He was omnivorous in his reading and his mind appeared to possess an inexhaustible capacity for storing information.

Having turned his back on the Aristotelian methodology which still dominated the university syllabus, he embarked enthusiastically on the works of currently fashionable thinkers like Descartes, Boyle and Hobbes. With equal enthusiasm he plunged into the study of the Bible and the early doctors of the Church. He took a close interest in current theological debates. For example, in 1690, he wrote *A Historical Account of Two Notable Corruptions of Scripture*. It was in this text that he added his weight to those who rejected the authority of the *comma Johanneum* (see p. 54).

The great plague visited Britain in 1665 and 1666. The university closed down and Newton went home. In isolation and free from other distractions, he applied himself to several problems of mathematics and physics. In what he later referred to as his golden period of discovery he brought to fruition or developed further several of the ideas he was already pondering.

The most famous incident that occurred at this time was the 'mystery of the falling apple'. It was a story Newton loved to tell and one he embellished over the years. It found its way into the first biography and from there into legend. Seated in his orchard, he explained, he saw an apple fall from its tree and wondered why free objects always fell *down*. In fact, he never answered the 'why' question; more important was the 'how' question. If the Earth was attracting the apple, that attraction must lie at its centre. Therefore, if the apple were free to make its way to the centre its course would describe a curve. This force must extend to everything on the planet – and beyond the planet. It must determine the motion of the moon – and the motion of the Earth in relation to the sun – and the movements of every heavenly body. 'Gravity', therefore, must be universal. If this was the case, it must be possible to calculate mathematically the mutual attraction of every star, planet and satellite and determine their sizes and relative distances.

Applying this principle involved mastery of Descartes' work in geometry that Newton had only recently begun to study. By a combination of observation and calculation, he discovered that 'the forces which keep the planets in their orbs must [be] reciprocally as the squares of their distances from the centres about which they revolve'.*

Twenty years would pass before Newton expatiated publicly on his revolutionary discoveries in *Philosophiae Naturalis Principia Mathematica* (*Mathematical Principles of Natural Philosophy*, 1687). Those years were replete with further intellectual accomplishments – and emotional traumas. In 1669, at the age of twenty-six, Newton became the Lucasian Professor of Mathematics. He had temporarily set aside his work on gravity to pursue his new passion, optics and constructed the first known reflecting telescope. A mere six inches in length, it was nonetheless more powerful than much larger refracting telescopes. It won him instant acclaim and an invitation to join the Royal Society. It also provoked what would be a long-running dispute with the society's curator of experiments, Robert Hooke.

* Letter to Pierre des Maizeaux (1718), Cambridge University Digital Library, MS Add.3968.41. Cf. R. Wagner and A. Briggs, *The Penultimate Curiosity*, OUP, Oxford, 2016, p. 269

Hooke (1635–1703) was a considerable scholar, the breadth of whose interests was scarcely less impressive than Newton's. He had made the vacuum pumps that enabled Boyle to demonstrate his law of the pressure and volume of gases. He had gone on to pioneer the development of microscopes and telescopes, developed new techniques for surveying and map-making, accurately charted the movements of Mars and Jupiter and proved the wave theory of light. In the early days of the Royal Society he was regarded as the leading English authority on optics. This by no means exhausted Hooke's speculations. He developed Hooke's Law of Elasticity which made possible the invention of the hairspring used in the first portable timepieces.

Just as climbers are drawn to mountains 'because they're there', so men of science cannot resist the puzzles presented by nature. Hooke also grappled with the phenomenon of combustion. Concluding that air contained a component essential for the process, he came close to the discovery of oxygen. Carrying out microscopic examination of human organs, he gave biology the word 'cell'. Hooke was also fascinated by palaeontology and tentatively put forward the opinion that certain fossils were the remains of creatures long since extinct ('tentatively' because theologians rejected the idea of life forms permitted by the Creator to disappear from the Earth). And much of this occupied his mind while he was working closely with Christopher Wren on rebuilding London after the Great Fire of 1666.

When Newton presented his telescope to the Royal Society and followed it up with a paper on the nature of white light, Hooke was underwhelmed and politely made his reservations known. This stung Newton into angry overreaction and a threat to resign from the society. The quarrel was patched up but, with both men, jealousy and mistrust simmered beneath the surface. Twenty years later, they boiled over again.

The two men had approached the mystery of gravity and planetary motions from different directions. Newton was a mathematical calculator. Hooke was an experimenter. The London-based employee of the Royal Society used the large internal spaces of St Paul's Cathedral and Westminster Abbey and a very deep well to observe the behaviour of long pendulums and the way suspended objects react at different distances from the planet's centre. From these observations he extracted geometrical data and arrived at conclusions about universal gravity.

Aware of Hooke's work (and, indeed, the work of others engaged with the same problem), Newton reignited his own interest. He made use of the records of John Flamsteed, the first Astronomer Royal who, during the course of his long professional life, used the observatory at Greenwich to catalogue more than three thousand stars and analyse the numberless movements of heavenly bodies. By this other route Newton reached similar conclusions to Hooke's, including the inverse square law of universal gravitation (see p. 246). But how similar were the two theories and who had got there first?

Newton was ready to publish the first part of his *Principia* in 1687. Hooke complained that his own work in the field had not been sufficiently acknowledged and that Newton was, in effect, guilty of plagiarism. Newton's response was to fly into a rage and excise from the manuscript all references to his rival. This unpleasant petulance became Newton's stock response to criticism. When he was preparing his second edition of the *Principia* around 1702–1703, he asked Flamsteed for access to the complete records of the observatory. Not wishing to be caught out twice, the astronomer refused. Newton's response was to use his influence at court to persuade Queen Anne's husband to authorise a royal star catalogue, thus obliging Flamsteed to make his data available. In 1708 the star project collapsed on the death of Prince George. Newton, however, had the last laugh. He was by now President of the Royal Society. He had Flamsteed expelled.

Distasteful as such academic bitchiness is, it cannot detract from Newton's achievement. The *Principia* went through several editions between 1687 and 1726 and was instantly recognised as the most important contribution made thus far to the understanding of the universe. There was, however, an aspect of the book's production that is seldom acknowledged. In 1693, the up-and-coming young classical scholar, Richard Bentley, had been designated to deliver the first series of lectures on Christianity and science for which provision had been made in Robert Boyle's will. He wrote to Newton for advice. The notoriously hermit-like scholar did not hesitate to respond:

> When I first wrote our treatise about our system, I had my eye upon such principles as might work with considering men for the belief of a deity and nothing can rejoice me more than to find it useful for that purpose.*

We do not have to take Newton at his own valuation. John Locke described him as a man whose knowledge of the Bible was almost unequalled. In his early student days he invested in copies of the sacred text to be distributed to the poor. Several people who knew him personally testified to his devotion and the genuine annoyance he expressed at the growing scepticism and materialism that were beginning to pervade fashionable society. But even without such testimony, the volume of his personal papers would leave no doubt about his genuine belief. Yet, this is the man who, in his last hours, refused to receive the Anglican sacrament. The story of Newton's

* Letter of Isaac Newton, 10 December 1692, Trinity College Library, 189.R.4.47. Cf. R. Wagner and A. Briggs, op. cit., p. 267

religious development is as fascinating as that of his scientific thought. Not that he would have drawn a distinction between the two.

> As the world, which to the naked eye exhibits the greatest variety of objects, appears very simple in its internal constitution when surveyed by a philosophic understanding . . . so it is in these visions [of God revealed in Scripture]. He is the God of order and not of confusion. And, therefore, as they that would understand the frame of the world must endeavour to reduce their knowledge to all possible simplicity, so it must be in seeking to understand these visions.*

This extract is from an unpublished treatise, just one of an immense archive of papers discovered after Newton's death. He was a compulsive writer and also a continuous reviser of what he wrote. His legacy of thousands of documents packed into scores of boxes remained unread until well into the twentieth century, when they found their way to the National Library of Israel. This archive is a two-edged sword for researchers. Potentially, it reveals the evolving pattern of his theological thought but that pattern has to be created from scraps of largely undated paper that render chronological analysis difficult.

However, certain significant events do stand out that enable us to suggest a few milestones in Newton's religious life. Around 1670, Newton had to contemplate ordination, which was required of all university fellows. He applied himself with his usual industry and incisiveness to the study of the Bible and the early Church Fathers. One result was that he found himself unable to accept the doctrine

* Newton, unpublished treatise on revelation, National Library of Israel, Jerusalem, Yahuda, NIS 1.1

of the Trinity. He concluded that the Church had gone astray at the Council of Nicaea (AD 325) when it formulated the creed which came to define Christian orthodoxy. Newton no longer believed in the divinity of Christ. Just as centuries of orthodox classical philosophy had been challenged – and challenged successfully – by post-Renaissance natural philosophers, so the time had come to reject the blind acceptance of theological orthodoxy.

According to his unpublished notes, he went further. Not only had the early doctors of the Church misunderstood Scripture, he claimed, they had actually been condemned in advance by Scripture. Newton made close study of the Bible's prophetic books, particularly Daniel and Revelation, and concluded that they foretold the emergence of Trinitarian error. His interest in prophecy brought him into the crowded arena of apocalyptic debate. His contribution was not to fix a date for the end time and the return of Christ in judgement. What he did propose as a result of his analysis of biblical prophecy was that this event would not occur 'before 2060'. These opinions should have obliged him to relinquish his fellowship. Fortunately, the Cambridge authorities, not wishing to lose a brilliant academic who was already making a mark in the wider world, decided to bend the rules in his favour.

As we have seen, the seventeenth century was a time of wide – not to say wild – speculation on all matters religious. Newton was, in common with all emancipated thinkers, obeying what he considered to be his responsibility to use his God-given reason in pursuit of truth. Just as he distanced himself from excitable preachers of imminent Armageddon, so he opposed natural philosophers who found in deism an answer to the mystery of Creation. God, he firmly asserted, was not a 'force' that could be equated with nature; a 'blind metaphysical necessity', as he protested in the *Principia*. He was one with Bacon, Galileo and others in asserting that the book of nature is a vital reference for those who want to explore the character of the

Creator. One of the qualities such study reveals, he proposed, is universal diversity. No depersonalised force locked within the created order could be responsible for such diversity; it would over and again produce the same answers. Only a being standing outside nature, originating it, directing it and intervening in it could produce the effects recognised by our senses. This intelligent and powerful being, Newton said, is what thinking men call the universal ruler, or in the Greek of the Book of Revelation, *Pantokrator*.

Newton hated Catholicism with a passion and the drift of events in the 1680s was sufficiently alarming to bring this hermit crab out of his shell. In 1684, Louis XIV revoked the eighty-six-year-old Edict of Nantes, which granted toleration to Protestants, and began a ruthless persecution of all opponents of the state Church. The following year James II succeeded his brother as king and began a vigorous campaign to undo the English Reformation. Newton ventured into university politics in order to resist attempts to place Catholics in positions of authority. In 1688, after James had been driven out in favour of his daughter Mary and her Protestant husband, William of Orange, he became one of the two university representatives in parliament. This involved him in protracted stays in London and it was here that he met Nicolas Fatio de Duillier.

This young Swiss scientist was a recent arrival who had made a sufficient impression on the learned community to be elected to the Royal Society. He became a passionate disciple of Newton and the attraction was mutual. The two men became very close and the famous Englishman paid his admirer to come and reside in Cambridge. Inevitably, some modern biographers have suggested that the relationship was homosexual. For that hypothesis there is no proof but what is clear – and this is no less sensational – is that the emotionally contained Newton developed a deep affection for the younger man. Fatio, for his part, embraced many of the religious and scientific ideas proposed by his mentor. He interposed himself on his friend's behalf in the dispute that blew up between Newton and Leibniz (see p. 254).

For his part, the older man allowed Fatio free access to all the papers he had hitherto kept secret from the world. It is no exaggeration to say that he opened his very soul to the young Swiss scholar. Was Fatio worthy of such devotion or was he simply using his charm to win the patronage of the great man and advance his own career? There is no doubting his ability. He taught mathematics. He delivered many papers to the Royal Society and contributed to learned periodicals. He made numerous astronomical calculations and carried out experiments aimed at improving the accuracy of timepieces. But his later career suggests a serious mental instability and this, presumably, lay at the root of a major crisis in Newton's life.

In 1693 the relationship came to an abrupt end and Newton had a nervous breakdown. He broke off relations with all his colleagues, declared his intention to resign from the Royal Society and made wild accusations against John Locke. Within a few years Fatio's mental energies were diverted into religious disputes and the study of alchemy and the cabbala. He became involved with the Camisards, an extreme group of French Protestants who took up arms – literally – against the Establishment after the revocation of the Edict of Nantes. A millenarian sub-cult was established in London and Fatio joined it. This led to him being accused of subversive activities and almost being torn to pieces by a London mob. After this he set off on an evangelistic European preaching tour. Fatio went on to live a long life (he died in 1753) but, although he attempted to re-establish his scientific career, he never regained his place at the centre of the academic community.

The break with Fatio probably prompted Newton to abandon his long-running study of alchemy. His unpublished notebooks contain more than fifty thousand words describing alchemical experiments. Yet he was not interested in the old dream of transmuting base metals into gold. His concerns were wider and deeper – how are chemical substances formed? How and why do they react with one another? What is the structure (what we would now call molecular structure) of metals? Newton's researches had involved him in, often clandestine, communication with pseudo-magical fraternities still interested in harnessing the elemental forces of nature to achieve power. It is not surprising that Newton would want to keep such connections secret. Fatio's enthusiasm for alchemy remained undiminished, as did his belief in the imminent end of the world. Newton's continued connection with this uncontrollable young man would only have damaged his reputation but the breaking of the relationship was nonetheless traumatic.

He recovered. More than that – he actually became sociable. In 1696 he was appointed Warden of the Royal Mint. He moved to London, with his niece Catherine Conduitt and her husband John. In 1703, after the death of Robert Hooke, he resumed his contact with the Royal Society and, within months, had been elected president. The change in his lifestyle can hardly have been greater. He had prestige and wealth through his position at the Mint. He now had the time and the resources to embark on his long-planned second edition of the *Principia*. One result of this was to reignite the old quarrel with Leibniz over the development of calculus.

The German scholar, like Newton, was now a venerable servant of government (in his case, the house of Hanover) who devoted the greater part of his energies to philosophical and mathematical problems. Like Newton, he also had a considerable following. The disagreement of the two men thus became a clash of two schools of thought suffused with nationalist emotions. In 1708 an article in the *Transactions of the Royal Society* gave official recognition to the old complaint of plagiarism rumbling round the scientific community. Leibniz took the initiative by demanding a retraction. This provided the opportunity for the showdown Newton wanted. He wrote a detailed report of the whole controversy, had it published under the aegis of the Royal Society and had it circulated as widely as possible throughout Europe. Not content with that, he wrote, anonymously, a review of his own work and included it in another edition of the *Transactions*. Leibniz followed suit by entering anonymous articles in various journals and encouraging his supporters to enter the fray. This unseemly bickering continued until Leibniz's death in 1716.

The literary battle extended to other topics on which the two principals disagreed. The most important lay in the realm of metaphysics. Theories about the composition of the universe and the behaviour of its component parts had implications for the existence (or non-existence) of a creator and the nature of that creator. Since Leibniz and Newton (despite his anti-trinitarianism) both believed in the Christian God, what they were debating was the character of this being.

Leibniz's Monadism (see p. 195), which he only proposed towards the end of his life, explained 'Life, the Universe and everything' in terms of elements (monads) that all came into being at Creation and will all cease to exist at the preordained end of all things. Monads do not die or change; they simply rearrange themselves (e.g. a prehistoric tree becomes coal because the monads react with each other in new ways). Everything is dependent on the will of God. Since He is perfect He only stands aside and allows the best of all possible machines to run. Man's role in all this was not to seek to understand the divine purpose (an impossibility) but to bow to the divine will.

Newton's concept of Creation was no less mechanistic. But he rejected the idea of God as an absent watchmaker who simply left the universe to run. The start point for Newton was not the *will* of God but the *rationality* of God. The elemental particles of the universe were lifeless; they could not move and interact without the constant organisation of the Creator. His God was one who intervened.

'What?' Leibniz responded. 'Was your God incapable of creating something perfect? Does he need to fiddle with it all the time?'

'Yes,' Newton replied. 'He is a caring God. If you don't believe in such a divinity, you're well on the way to atheism.'

In Newton's mind gravity was the clearest proof available to date of the existence of this intervening and caring God. Without this mysterious, external force the universe would long since have imploded. Gravity was awe-inspiring, mathematically beautiful – and quite incomprehensible. Incomprehensible, that is, in terms of natural philosophy.

As Europe stood on the doorstep of the Enlightenment, some thinkers found themselves looking at scientific and religious doors to different rooms, marked, respectively, 'How?' and 'Why?'

CHAPTER TWELVE

Avant le deluge

In 1747 an amateur theologian received an LLD degree from Oxford University for a dissertation, subsequently published as *Observations on the History and Evidences of the Resurrection of Jesus Christ.* His name was Gilbert West and his book ran to several editions. In his Preface he wrote:

> I am not ignorant how little reputation is to be gained by writing on the side of Christianity, which by many people is regarded as a superstitious fable not worth the thought of a wise man; and considered by more as a mere political scheme adumbrated to serve the power and interest of the clergy only.*

The words offer a rough guide to prevailing religious attitudes in mid-eighteenth-century Britain. There were those who believed in the truths of Christianity; those who denied those truths; and those who espoused a formal, 'priestified' religion. We seem to have come a long way from the world of Locke, Leibniz and Newton.

In the philosophical debates of the turn of the century there had been sceptics. There had been thinkers whose religious opinions had strayed some way from Christian orthodoxy. There were deists who acknowledged no 'revelation' beyond that which came by way of reason. But there had been very few out-and-out atheists (or, at least, very few prepared to nail their colours to the mast). But by the middle of the century the gap between religion and science was manifestly

* G. West, *Observations on the History and Evidences of the Resurrection of Jesus Christ* (1749), fourth ed., Gale Ecco, Michigan, 2010, 'Preface', p. xii

wider. Change was not obvious merely in the intellectual world of the fashionable salon and the university classroom. The implications of philosophical and theological theory were impacting on practical politics and economics. Society was on the move.

Louis XV, or it may have been his mistress, Madame de Pompadour, is credited with the prophetic observation, in 1757, '*Aprés moi le deluge*'. The cataclysmic collapse of the Ancien Régime in the revolution of 1789 would certainly be no spontaneous, out-of-the-blue explosion of public anger. Britain's problems with her American colonists, which had begun two decades earlier, had raised issues of 'liberty, equality and fraternity' before those principles became a political mantra. History is a continuum of sudden events, a chain of causes and effects. What philosophers were writing and preachers were preaching were part of the build-up to the dynamic changes in society throughout Europe and, increasingly, in Europe's overseas possessions.

The changes of the eighteenth century also bear witness to the cyclic nature of human history. We saw how the scepticism of the Renaissance led not only to the challenging of political and religious establishments and the quest for other sources of wisdom, but also to the reaffirmation of fundamental Christian beliefs in new, dynamic ways. This pattern now repeated itself. Viewed from our vantage point the pre-revolutionary era has a Janus-like aspect. One face is represented by secularist philosophy. The other wears the earnest gaze of religious revival.

One critic of traditional Christian belief Gilbert West had in mind was David Hume (1711–1776). He was certainly not alone in disapproving of this freethinking Scot. Hume was recognised as a man of impressive intellect but his applications for professorial chairs at Glasgow and Edinburgh were both turned down. The academic authorities did not fully understand the Lowland philosopher but what they did understand they did not like. Theirs was a valuation that would be echoed by many others during the following centuries.

David Hume was the younger son of a well-connected but not affluent family. His father died when the lad was two. He attended Edinburgh University but did not graduate, was put to the study of law and abandoned it in favour of philosophy. Being left very much to his own devices, he did a variety of jobs to earn a living but it was the world of books that enchanted him. He was a voracious reader and a compulsive writer. In an autobiographical work written towards the end of his life he confessed that his ruling passion had always been the desire for literary fame, an admission that goes a long way towards explaining his long career and the – sometimes contradictory – ideas that he proposed. To be accepted by the literati he had to cultivate the leaders of intellectual fashion while at the same time drawing attention to himself as the voice of original – not to say outrageous – ideas.

For a Lowland Scot the mid-eighteenth century was not the easiest time to pursue will-o'-the-wisp celebrity. Dr Johnson may have had Hume in mind when he famously observed that 'the view of the London road was the prospect in which every Scotsman most naturally and most rationally delights'. It was not only Highlanders who nursed ambitions of restored Stuart rule. Several of Hume's friends were Jacobites and he was very careful to keep his own counsel on the vexed political and social debates pre-occupying thinking men on both sides of the border. But his prospects and his inclinations lay in seeking patronage within the dominant English Tory establishment. He even changed the spelling of his name from 'Home' to 'Hume' in order to make it easy for the English to pronounce it.

His life falls, fairly distinctly, into two parts: before and after 1761. His earlier years were spent in profound and extensive study (which, at one point, resulted in a mental breakdown). To this period belong most of his philosophical writings, to which we will return shortly. But his grasp of ancient and modern philosophy and his own forthright, angry-young-man opinions failed to impress the eighteenth-century intelligentsia. Eventually, he got the message: he had to choose between trumpeting his views on 'life, the universe and everything' and achieving fame and fortune by some other means.

In 1752 he was appointed keeper of the library of the Faculty of Advocates in Edinburgh. This gave him free access to thirty thousand volumes. He fell to study with his usual frantic energy and employed his now well-honed literary talents to writing a history of Britain. The work eventually ran to six volumes, appearing between 1754 and 1761 and, after a shaky start, was acclaimed as the best treatment of national history yet written. By the latter date he had become, as he himself said, 'Not only independent but opulent'.

From this point Hume's career as a philosopher virtually ceased. He had become a celebrity, able to mix with the elite of British society. He was appointed to various government posts at home and abroad, eventually rising to the position of Under-Secretary of State for the Northern Department, the government body responsible for relations with the Protestant states of northern Europe. It was while serving as secretary to the British Ambassador in Paris (1763–1765) that he became friendly with the elderly French radical, Jean-Jacques Rousseau, a controversial writer who had made many enemies. Hume offered Rousseau refuge in England and did all he could to provide the exile with a peaceful environment in which to continue his work. The relationship ended badly with a paranoid Rousseau turning against his saviour (see p. 271).

Once he 'arrived', Hume basked in his celebrity. He became the very apostle of fashionable scepticism. In London and in France he had his acolytes. When back in Edinburgh he attended the weekly meetings of the Select Society (which later became the Poker Club) and discussed matters philosophical with other leading intellectuals. Socially, he appeared as an affable bon viveur, a veritable Epicurean.

Sometimes, it seems, he had passed beyond scepticism to a detached cynicism. For example, when a young non-believer considering the possibility of taking Anglican orders asked Hume's advice on what was obviously for him a moral issue, he received the reply that he should go ahead and not trouble himself with the religious peccadilloes of the vulgar majority.

Hume's writing also reveals a dislocation of style and substance. By the time he came to write his *History*, he had acquired a fluency and expressiveness (partially gained from his familiarity with French literature) that outdid his rivals and made his account the standard work in its field. It was, however, beset with factual errors and the misconceptions of archival records. But the author seldom troubled himself with detailed research. He aired his own prejudices – but only when he calculated that they would not have a deleterious effect on sales. Thus he wore his Toryism on his sleeve but was careful in his comments on Jacobitism and atheism. With the passing of the years and the development of more scientific assessment of primary sources, Hume's historical work fell out of favour. At the same time his earlier, philosophical writings received greater attention.

In his reflections on 'life, the universe and everything' Hume started, like many thinkers since the Reformation, with an aversion to the Church as he experienced it. The authoritarianism of the Catholic hierarchy, no less than its Anglican and Presbyterian counterparts, irritated him. Enthusiasm he could not abide. In *An Enquiry Concerning Human Understanding* (1748) the author sought to pull the rug from under all Christian theology by means of an argument that would be 'an everlasting check to all kinds of superstition delusion'. His argument (by no means wholly original) was to challenge belief in miracles. His empirical method was not examining any particular 'unnatural' event and providing a rational explanation for it but the assertion of principles which, he claimed, would apply to all such events. He then proceeded to define 'miracle' in such a way as to render it self-evidently non-existent:

> A miracle is a violation of the laws of nature; and, as a firm and unalterable experience has established these laws, the proof against a miracle, from the very nature of the fact, is as entire as any argument from experience can possibly be imagined.*

Hume went on to argue that the transmission of information about miracles was suspect. First-hand experience is almost non-existent; we rely on testimony passed on via several raconteurs. Moreover, such tales were relayed for the most part by members of 'ignorant and barbarous nations'.

As to the existence of God, Hume's first line of attack was upon the theory of cause and effect. While the human mind is capable of discerning *related* events, he insisted, it could not rationally demonstrate that one was the necessary antecedent of another. It was, therefore, impossible to postulate logically the existence of a first cause. Hume had his own theory about the origin of monotheism. He traced the history of religion from polytheistic nature worship, through a series of contracting beliefs, to the point at which all power and virtue were seen to reside in one deity. In concluding his argument, he could not resist a final sally against the prevailing monotheistic systems:

> What so pure as some of the morals, included in some theological systems? What so corrupt as some of the practices to which these systems give rise?†

* D. Hume, *Enquiries Concerning Human Understanding and Concerning the Principles of Morals*, L. A. Selby-Bigge, ed., OUP, Oxford, 1975, pp. 130–1
† D. Hume, *The Natural History of Religion*, H. E. Root, ed., A&C Black, London, 1956, p. 76

If we were to judge from Hume's various written observations about religion, we would have to conclude that he quite unmistakeably threw down the gauntlet to all systems reliant on revelation. Later advocates of atheism have not hesitated to claim him as a founding father. Yet he was far from being an angry and defiant challenger of conventional Christianity. His amiable disposition and his skilful courting of members of the Establishment protected him from an angry backlash that could have, at best, seen him outlawed from polite society. If he calculated carefully the risks he might take, it was probably with the fate of La Mettrie (1709–1751) in mind.

In the very year, 1748, that he published his *Enquiry Concerning Human Understanding,* Julien Offray de La Mettrie fled from Leiden to Berlin (having three years earlier fled from Paris to Leiden). This talented physician had, as a young man, contemplated a career in the priesthood but subsequently turned violently against the Church. In his writings he proposed that man was merely a machine or a superior animal and advocated not only atheism, but hedonistic materialism. Such extremism shocked even radicals such as Voltaire and Diderot, who were, doubtless, not surprised to learn of La Mettrie's death from gluttonous overindulgence.

Hume's death in 1776 prompted this entry in Boswell's *Life of Dr Samuel Johnson*:

> I mentioned to Dr Johnson that David Hume's persisting in his infidelity, when he was dying, shocked me much. JOHNSON: 'Why should it shock you, sir? Hume owned that he had never read the New Testament with attention. Here, then, was a man who had been at no pains to enquire into the truth of religion and had continually turned his mind the other way. It was not to be expected that the prospect of death

would alter his way of thinking, unless God should send an angel to set him right."*

The first contemporary philosopher who *did* make a close study of the Christian scriptures and put forward a rational refutation was Hermann Reimarus (1694–1768), a teacher of oriental languages in Hamburg. Like other freethinkers, he was careful not to go into print with his more extreme ideas. His most famous (or notorious) treatise, *Apology, Or Defence of the Rational Worshippers of God,* remained in manuscript at the time of his death. A member of his intellectual circle was Gotthold Lessing (1729–1781), an up-and-coming dramatist and critic. He it was who inherited Reimarus' papers and it was he who published the *Apology* in sections over the next few years. Their impact on Christian theology was profound.

What Lessing/Reimarus did was apply to the Bible the standards of rational critique that would be applied to any other collection of ancient texts. The Christian writings, he asserted, were not infallible. Ancient miracles without modern standards of proof could not be used in support of metaphysical truth. This barrier to religious conviction was what Lessing called 'Lessing's Ditch'. Neither he nor Reimarus wished, like Hume, to consign Christianity to the dustbin of broken and discarded myths. Indeed, what they aimed to do was reinterpret the ancient faith in such a way as to make it intellectually respectable.

How far they were prepared to go was indicated in *Apology,* in the section published as the last chapter of the *Fragments of an Unnamed Author.* It was entitled *On the Intentions of Jesus and the Evangelists.* This elaborate reworking of the Gospels and the Acts of the Apostles presented Jesus as a Jewish political leader who had tried to reform and re-energise the chosen people in preparation for the establishment of

* J. Boswell, *The Life of Samuel Johnson* (abridged), Dent, London, 1909, p. 409

the Kingdom of God on Earth. His mission had failed and resulted in his death. After the crucifixion the apostles had invented the story of the resurrection and held out the promise of their leader's final return in triumph. Unlikely as this scenario was, it provoked an angry response from orthodox Christian writers. The result was a feverish pamphlet warfare that only came to an end when a hastily drafted new law imposed censorship on unorthodox religious publications.

Thereafter, Lessing restricted his talents largely to drama, poetry and literary criticism. But these gave him plenty of scope for philosophical comment on human affairs. What, above all, he championed was freedom – freedom from traditional stylistic restraints in the arts; freedom from imposed dogma, whether religious or political; freedom from economic control by the nobility; and freedom from tyranny – in fact, freedom for all human beings to develop their potential, unencumbered by traditional beliefs and the social structures buttressed by those beliefs. It was an ambitious manifesto but one more and more people were coming to share. Philosophy was no longer confined to the rarefied atmosphere of the lecture hall and the salon but was infiltrating the stage, the novel, satirical prose and popular journalism. The view from the windows of the Louvre was one dominated by gathering dark clouds. The deluge was, manifestly, on the way. And it was in France that it was most visible.

François-Marie Arouet (1694–1778) was a rebel. Born into a wealthy, landed family, he refused to use his patronymic and styled himself 'Voltaire'. Reared by Jesuits, he kicked over the traces and gave himself to the pleasures of the flesh. Put to the law by his father, he abandoned his studies in order to pursue a literary career. Rejecting the social mores of his class, he specialised in biting satire. In his early years his outspoken criticism earned him spells in prison and periods of exile. It was during a sojourn in England between 1726 and 1729 that he learned to ally his negative cynicism with the empiricism of Locke and the scientific methodology of Newton. It was Newton, he affirmed, who taught him to 'examine, weigh, calculate and measure, but never conjecture'.*

He did his best in his *Elements of the Philosophy of Newton* of 1738 to enlighten his fellow countrymen on the work of his English hero. With age came a modicum of discretion which, coupled with his literary brilliance, won him election to the Académie Française. In 1750, Voltaire, like La Mettrie and other freethinkers, was drawn to the cultured court of Frederick the Great in Berlin, where he received a royal appointment. Banned from Paris by Louis XV, he spent the years 1755–1778 at first in Geneva – where, unsurprisingly, he found the prevailing Calvinism not to his taste – and then on a French country estate at Ferney, close to the Swiss border. It was in this period that he wrote his most influential works. His satirical play, *Candide or The Optimist* (see p. 194) appeared in 1759. His conclusion, that human well-being consists of simply making life as agreeable as possible, was not far removed from Hume's Epicureanism.

But Voltaire was no atheist. He despised rejection of belief in God and he did so largely on practical grounds. In his assessment of human character he bracketed together atheism with religious fanaticism; both were 'almost always fatal to virtue'. Like other thinkers we have encountered, the principal target of Voltaire's wrath was organised religion. His last major work was a collection of essays, most of which attacked aspects of the Church and its teaching. The *Philosophical Dictionary* was published anonymously in 1764 and was enlarged in subsequent editions. Its importance may be gauged by the fact that it sold out and that, in many places, copies were seized and burned by the authorities. Mixed reactions followed him to the end. When he finally returned to Paris in 1778 it was as a public hero. When he died he was refused Christian burial.

* Cf. R. R. Palmer, *The Age of Democratic Revolution: The Challenge*, Princeton University Press, New Jersey, 1959, p. 214

As to his positive beliefs, they may be described as a form of deism. The splendour, immensity and intricacy of the cosmos put the existence of a creator beyond doubt. Voltaire venerated Christ as a great teacher – on a par with Socrates and other founders of Western civilisation. The tragedy, in his eyes, was that seventeen centuries of Church dogma and abuse of power had distorted his wisdom beyond recognition.

Most members of the avant garde would have agreed but that does not mean that we can think of the Enlightenment as a movement with clear aims and an agreed agenda. It was more of a *mood* projected by various promoters, each with his own interpretation. In this it was like the Reformation: its leading spirits were united in what they were against but far from united in what they were for. It would be a mistake to look back across the crises of the French and American revolutions and think that Enlightenment was setting out the blueprint for rebellion, social equality and democracy. For example, one fellow radical whom Voltaire heartily disliked was Jean-Jacques Rousseau.

Nobody much liked Rousseau (1712–1778). Indeed, he was a rather unlikeable man. His difficult persona is partly explained by his early history. Born in Geneva, he never knew his mother, who died as a result of natal complications, and he saw little of his father, who moved away when the boy was ten. His subsequent education was overseen at first by a Calvinist minister and then by a Catholic noblewoman devoted to the cause of converting Protestants. Subsequently, he became the lover of this lady, who was thirteen years his senior.

Having unsuccessfully tried various avocations, Rousseau settled to serious study in his twenties and, in 1743, secured the post as secretary to the French ambassador in Venice. His sojourn there was brief but it did deepen his love of music, for which he had a natural aptitude. We next find him in Paris, where he wrote an opera which was performed before the royal court. Louis XV was sufficiently impressed to offer the writer a pension. But Rousseau was not impressed with the king and rejected the patronage. He was now living a bohemian existence, partnered with an illiterate seamstress, who presented him with several children, all of whom were placed in a foundling hospital. His difficulty in making and sustaining relationships seems to have been rooted in personal insecurity (hardly surprising, given his upbringing) that gradually evolved into a persecution complex.

It was in Paris that Rousseau became involved with the *Encyclopédie*:

> Encyclopedia – 'the circle of teachings' – may be taken as the emblem of the eighteenth century. Like the Renaissance, the age was confident that the new knowledge, the fullness of knowledge was within its grasp and was a means of EMANCIPATION ... Everything will ultimately be known and 'encircled'.*

The excitement created by a growing treasury of scientific discoveries spread from the intelligentsia to all levels of literate society. People craved information about the latest developments and the implications of those theories for everyday living. Encyclopedias were a way of satisfying that demand. Compendia of knowledge were not new; ever since ancient times scholars had produced lists and dictionaries of various kinds – herbals, bestiaries and cosmologies as well as more general collections – but in the eighteenth century the encyclopedia attained a new level of comprehensiveness – and influence.

This phenomenon was extensive both because of the sheer amount of information to be communicated and also because of the voracity of the audience. Literacy was spreading. People were reading newspapers (in Britain by 1720 there were twelve published in London and

* J. Barzun, op. cit., p. 359

twenty-four in the provinces). The circulating library had been born. An increasing number of people lived in towns and cities and could gather in clubs, coffee houses and inns to share the latest news. The Church, proud patron of education at various social levels, was losing its grip on the training of minds (the Jesuit order came under increasing attack and was finally abolished by papal decree in 1773). The dissemination of information was becoming secularised. In 1728 Ephraim Chambers published *A Universal Dictionary of Arts and Sciences – Explaining not only the Terms of Art but the Arts Themselves.* It was this that inspired a French counterpart.

The *Encyclopédie* was launched in 1751 'to collect all the knowledge that now lies scattered over the face of the earth, to make known its general structure to the men among whom we live and to transmit it to those who will come after us'. But the new publication (extended by 1780 to thirty-five volumes, including index) was not just an accumulation of factual information. Its originators aimed to make readers 'more virtuous and more happy'. In other words, the *Encyclopédie* would bulldoze the rubble of outmoded ideas and beliefs and build a temple where men might worship new truth – truth that would set them free.

The principal architect of this publishing skyscraper was Denis Diderot (1713–1784). In what was becoming something of a pattern with new thinkers, he toyed with the possibilities of a career in the Church or law before deciding to become a writer. In all aspects of his life Diderot was industriousness personified. Between 1745 and 1782, thirty-seven books and pamphlets came from his pen. He maintained a mini-harem of mistresses and turned some of his sexual exploits into bawdy prose. But it was his dedication to the *Encyclopédie* that was truly prodigious. Having spent several uncomfortable months in prison, he resolved to turn his back on his more scandalous pursuits and concentrate on the enterprise for which he had been appointed co-editor at the end of 1747. He worked day and night writing material, recruiting other writers, editing text and negotiating with printers. When the first volumes were ready they were, inevitably, very expensive. The *Encyclopédie, ou Dictionnaire Raisonné des Sciences, des Arts, et des Métiers par une Société des Gens de Lettres* (*Encyclopedia, or a Systematic Dictionary of the Sciences, Arts, and Crafts*) was a financial risk. But it paid off. The first printing sold quickly and subsequent volumes also did well. The rapid success of the *Encyclopédie* is reminiscent of the reception accorded to Luther's *Ninety-Five Theses*. It, too, was an idea whose time had come. Like the reformer's tract, it too raised a storm of protest; and, like the reformer's tract, this, too, flourished on opposition.

One of the contributors Diderot recruited was Rousseau, who produced several articles, notably some detailed and trenchant observations on contemporary music. He and Diderot agreed that, to live the enlightened life, reason was not enough. But its accompaniment was not, as other philosophers proposed, revelation. It was passion. Man, they believed, was a feeling as well as a thinking animal and happiness involved keeping these two horses running smoothly together in harness.

But Rousseau was incapable of making common cause with anyone else for long. By 1754 he had fallen out with Diderot, who accused him of plagiarism and excoriated him as 'false, vain as Satan, ungrateful, cruel, hypocritical and wicked'. By this point Rousseau had returned to Geneva and reconverted to Calvinism (rejection of atheism was one cause of his estrangement from Diderot). When his writings continued to outrage the authorities, he followed the route taken by other radicals in seeking the protection of Frederick the Great. The Prussian king was ruler of the canton of Neuchâtel and Rousseau lodged there until the local authorities expelled him. But notoriety in official circles inevitably aroused admiration among the radically inclined. In his wanderings Rousseau seldom lacked for offers of help from those who admired his outspoken criticisms of religious and social conventions. It was in January 1766 that he accepted David Hume's offer of refuge in England.

Friends warned the Scot that there would be 'tears before bedtime'. He would have done well to listen. There were two indicators of trouble ahead: Rousseau's instability and his fame. Much of the Swiss radical's popularity rested on his two novels, *Julie, Or The New Heloise* (1761) and *Emile, Or On Education* (1762). The author had found the knack of enshrining his philosophy in prose that was both elegant and passionate. By the time of his arrival in England he had an international following. The press closely watched his every movement. Correspondents in all the European capitals plied their English contacts with eager demands for the latest news of the controversial celebrity. Whatever happened during his English sojourn would be widely known within days.

Was Hume jealous of the attention his guest was receiving? Was the sensitive Rousseau on the lookout for the merest suggestion of a slight? Perhaps both. Within months harsh words were spoken and written. Rousseau complained of a plot to dishonour him. Hume was incensed at his guest's ingratitude. Popular magazines throughout Europe were soon sensationalising the feud. After sixteen months Rousseau recrossed the Channel, shaking the dust from his shoes against England.

What this episode illustrates is that the 'Enlightenment' was not a clearly defined movement. It is even less precise than 'Renaissance' and 'Reformation' and, heaven knows, those terms were fuzzy-edged enough.

> Hume was a combination of reason, doubt and scepticism. Rousseau was a creature of feeling, alienation, imagination and certainty . . . Hume was an optimist, Rousseau a pessimist; Hume gregarious, Rousseau a loner. Hume was disposed to compromise,

> Rousseau to confrontation ... Rousseau revelled in paradox; Hume revered clarity. Rousseau's language was pyrotechnical and emotional, Hume's straightforward and dispassionate.*

If we look for similar examples of bitter clashes between men divided by common conviction, we might instance Luther and Zwingli or Calvin and Servetus. Rousseau's major heresy from a rationalist viewpoint was his belief in and his emotional commitment to God. His religion was, it goes without saying, far from orthodox (whether Catholic or Reformed) but it was certainly not the deist's acknowledgement of a distant, uninvolved watchmaker. He believed in toleration, refusing to accept that the monotheistic faiths 'owned' God. By the same token he rejected atheism. For him the wonders of nature demanded a response of humility and awe; to him the intellectual dissection of both rationalists and theologians was repugnant. In *Emile* he wrote, 'The best use I can make of my reason is to resign it before thee [i.e. God]; my mind delights, my weakness rejoices to feel myself overwhelmed by thy greatness'.†

When we come to consider Rousseau's political philosophy, it is important to distinguish between what he believed and what people believed he believed. Contemporary and later idealists, influenced by the popular image of the *enfant terrible* of the Enlightenment, quoted his colourful language (frequently out of context) to support their own ideologies. Rousseau began his *Social Contract* (1762) with the ringing proclamation, 'Men are born free, yet everywhere are in chains'. Was he sounding a clarion call to revolution? Certainly not. He went on to analyse the chains holding together civil society and

* D. Edmunds and J. Eidinow, 'Enlightened Enemies', *The Guardian*, 26 April 2006. For a fuller account see the same authors' *Rousseau's Dog: Two Great Thinkers at War in the Age of Enlightenment*, Faber, London, 2006

† J. J. Rousseau, *Emile*, trs. B. Foxley, Dent, London, 1921, pp. 248–9

the conditions that legitimated their existence. He was continuing the debate in which Hobbes, Locke and others had been involved.

Like earlier writers, he considered the various kinds of polity men might live under – monarchy, aristocracy and democracy – and proposed that power derives from and should be exercised by the sovereign. But this 'sovereign' is not Hobbes' leviathan, an individual to whom the members of the state yield their power. It is the people who are sovereign. The actual type of government pertaining in a state was less important than the establishment of a mechanism ensuring that the will of the people was expressed and made the basis of law. Executive power had to exist and had to be enforced but only insofar as it was acceptable to the people. This involved a system of checks and balances. In the ensuing century the concept of the 'consent of the governed' would be stated, denied, restated, modified and hammered out in state constitutions in Europe, North America and wherever democracy was prized. But, when Rousseau wrote, these principles were by no means acceptable to or understood by the rulers or the majority of the ruled.

When the Reform Act of 1867 extended the British franchise, Viscount Sherborne would sourly remark, 'We must educate our masters.' A century earlier Rousseau agreed with the sentiment, if not with the spirit in which it was expressed. The sovereign people in his ideal society would be reared in such a way as to exercise their responsibilities conscientiously. In *Emile* he attacked prevailing attitudes towards the rearing of children and, by implication, the structure of society for which each rising generation was being moulded. The individual was key. The child must not be forced, by birch and precept, to fit into a predetermined place in the existing order. The central character of Rousseau's tale is reared in rural isolation, away from the corrupting influences of church and state, in order to appreciate and be at one with the wonders and beauties of nature. The infant Emile is not a blank sheet on which society's beliefs and principles must be impressed, but someone with unique gifts who had to be nurtured. As far as religious upbringing was concerned, no dogma, whether Catholic, Protestant, deist or atheist, was imposed on Emile, who, therefore, grew up to embrace 'natural' religion.

Rousseau's pedagogic principles enraged doctrinaire advocates of all schools of thought, though, in fact, they resonated with one principle basic to most shades of Protestant theology, namely their stress on the individual's relationship with God. Salvation involved a conscious response of will to what each person perceived of the God revealed in Creation and Holy Writ. His/her spiritual development was dependent, primarily, on the 'daily walk' with God and only secondarily upon the theology inculcated by pastors and preachers. This old belief reasserted itself among Christians with fresh fervour and power in the age of Enlightenment and was, in some ways, connected with the Enlightenment itself. It is to the phenomenon of revivalism that we must now turn.

As we have already observed, there was nothing new about scepticism towards the Church. It was shared by thinkers who rejected *all* conventional religion and convinced Christians ardently seeking a dynamic faith. Among the latter the passionate desire for a more affective religion produced the mass phenomenon called 'revival' – a movement that leaped across the Atlantic:

> America, like England, witnessed a resurgence of religious passion, which put forward old ideas: consciousness of sin and recognition of God's mercy; self-reform imperative to ensure grace and salvation.*

The old question, 'What must I do to be saved?' still refused to go away. This surge of spiritual awareness was not confined to the

* J. Barzum, op. cit., p. 404

English-speaking world. In Lutheran Germany and Scandinavia the complementary movement was Pietism. Here, as elsewhere, one of the tap roots of the movement was mysticism, which was fed by Christian contemplatives like Jacob Boehme (1575–1624), Thomas à Kempis, medieval ascetics and even Paracelsus, as well as philosophers like Pascal who had stressed the inadequacy of rationalism in man's quest for God. In England it was writers like William Law (see p.279) and John Bunyan from whom readers learned about the 'inward' religion of divine grace and its operation in the life of the believer.

The various aspects of this movement crossed, merged, separated, faltered and revived over the course of the century and it is impossible to do justice to them all in the space available here. However, if we seek a 'significant' date as our starting point we could with some justification advance the claims of 1727. In this year when Isaac Newton died, three other events occurred which few, if any, contemporaries would have thought worth recording. In Oxford, young Charles Wesley started what fellow students sneeringly called the 'Holy Club'. In Prussia, King Frederick William I threw the weight of the state behind the Pietist movement. In Saxony, the 'Moravian Covenant for Christian Living' was signed by members of Count Zinzendorf's new community at Herrnhut. What critics deplored as overblown religious emotionalism – 'enthusiasm' – merged with core Protestant theology as evangelicalism. This movement now began a determined infiltration of society.

In Prussia, as in much of northern Europe, there had been many who pressed for further reform ever since the establishment of Lutheranism. They had reservations about the degree of toleration afforded to old, Catholic dogmas and practices by the leaders of state Protestant churches. However, any theological disputation had been pushed into the background by the devastation of the Thirty Years' War. Not only had the conflict destroyed lives and property, it had disrupted society, including education and regular religious ritual. Church leaders complained that their flocks 'no longer knew who Christ or the Devil were'.* There were also those who did not even care who Christ or the devil were. Public mores, if we are to believe the protesters, were dominated on the one hand by libertinism and on the other by chiliasm. While worldlings were living as though there was no tomorrow, fanatical preachers were warning that there *was* a tomorrow – the day of God's fearsome judgement. It was against this background that a minority element – Pietists – tried to promote a religion of inner renewal and personal sanctification that would, along with other benefits, bring about the salvation of society.

The leading figure in the early Pietist movement was Philipp Spener (1635–1705). His demanding teaching of moral perfection and spiritual discipline challenged conventional Lutheranism and won him many enemies but in Frederick II of Brandenburg (who also reigned from 1701 as Frederick I of Prussia) he found a firm supporter. Frederick was busy raising the profile of his regime as a major military power but was also a supporter of the arts and sciences. He founded the University of Halle and it was here, between 1694 and 1705, that Spener exerted considerable influence over the student body. He passed on the torch to his disciple Auguste Francke (1665–1727), who became Professor of Theology at Halle.

Francke was a veritable spiritual bulldozer. He continued the battle against the Lutheran establishment. Through preaching and education he built up a cadre of pastors and teachers to spread a religion based on personal experience of the divine through scripture and the 'inner voice'. He founded schools, an orphanage and a student hostel. Other related ventures included a printworks for

* Cf. M. Fulbrook, *Piety and Politics – Religion and the Rise of Absolutism in England, Württemberg and Prussia*, CUP, Cambridge, 1983, p. 24

Bibles and tracts, and a colportage system for distribution of this improving literature. This would all have been impossible without the support of the government. Frederick William I, the 'Soldier King' (reigned 1713–1740), was as industrious in his politico-military sphere as Francke was in the religious one. He was building a powerful modern state with a strong army and an efficient bureaucracy. Though himself a Calvinist, he saw the value of the well-organised and disciplined Church Francke and his team were building. As a young man he had experienced a profound religious conversion and, although this did not manifest itself in a life of pious, restrained devotion, it did mean that he and Francke understood each other. The king appreciated the clear teaching and disciplined regimen of the Pietists and increasingly supported them against their critics. By the time of Francke's death in 1727, the Pietist bandwagon was rolling steadily and the king ensured that his place was filled by others able to continue realising his own vision. He decreed that all candidates for Church office must study theology at Halle and appointed Pietists to leading positions in Church and state.

One of the students at Halle was Nikolaus Ludwig von Zinzendorf (1700–1760), a prosperous landowner of ancient family. He took his territorial responsibilities very seriously but also wanted to create an environment in which he and others could develop spiritually and breathe some life into the local Lutheran church. This he did by preaching and publishing Bibles and tracts. His little Pietist community grew and began to attract refugees from religious persecution in other areas. Zinzendorf settled the newcomers in the village of Hernhutt, on his Saxon estate. What developed was a lay community where Christians of different denominational backgrounds could live together bound by the Moravian Covenant for Christian Living. The name 'Moravian' derived from the region of Bohemia (now part of the Czech Republic) from where the founders had originally fled.

The emphasis was on devotional exercise and practical service, rather than doctrinal niceties.

Missionary endeavour developed naturally within this nucleus. Beginning in the 1730s, groups of Moravians were sent out, first of all to other parts of Germany and the Baltic lands, but soon further afield. This activity was made easier by movement of European peoples to colonies and trading settlements overseas in this era of expansion. The dynamic of Zinzendorf's movement is demonstrated by the geographical extent of its missionary activity. Within twenty years Moravians were working in most of the North American colonies, Labrador, the West Indies, South America, the East Indies, Egypt and South Africa. There were many reasons for this success, not least of which was the training of Zinzendorf's envoys. The emphasis of their preaching was always on religion of the heart, rather than of the head. This did not mean that their theology was non-rational, simply that its focus was on the work of the Holy Spirit within the believer, rather than on understanding of theological concepts. The study of the Bible and devotional works had an important place in the life of the believer as he/she grew in spiritual maturity. But initially the simple Gospel was couched in terms that people of very different backgrounds could relate to.

In England the medley of sects and cults that appeared in the middle years of the previous century had also indicated a rejection of the formal ritual of the state Church and a quest for a more personal, independent religion of the heart rather than the head. But after the Restoration all forms of dissent were banned. Some withered under persecution but, as usually happens, others were driven to the margins of society where they drew strength from their sufferings. The spectrum of dissent ranged from quietists, such as Quakers, through Baptists, Congregationalists (Independents) and Presbyterians to Unitarians. This non-conforming community proved too strong to be squeezed out of existence.

One key area in which the separatists mounted an important challenge was education. The English universities were closed to them. Their response, as we have seen, was to set up their own dissenting academies, staffed, in large measure, by the clergy who had been deprived of their livings. By provoking this independent educational movement the government had shot itself in the foot, for the academies were able to teach their own version of Christian theology to new generations of ministers, preachers and laymen.

These institutions, inevitably, saw themselves as being in competition with Oxford and Cambridge. Some claimed, justifiably, to be more in touch than the ancient seats of learning with the latest developments in philosophical and theological thinking. They taught Newtonian physics and were open to Enlightenment ideas wafting across from the Continent. 'Their philosophy was an active preparation for a new age . . . they formed a great, permanent undercurrent of dissatisfied criticism of the state of England.'* From the ranks of these academies came some of the leading businessmen and *hommes d'affaires* of the eighteenth century.

Yet it was from within the universities and the established Church that the most dynamic religious movements of the age emerged. William Law (1686–1761) provides a useful link between old elements of affective religion and the revivalism of the eighteenth century. Law was no dissenter – far from it. He was a fellow of Emmanuel College, Cambridge, a High Church Anglican, committed to the close connection of Church and state. Where he parted company with the Establishment was over the Protestant succession. Refusing the oath of allegiance to George I cost him his Cambridge fellowship in 1714. After some years spent partly as tutor in the household of Edward Gibbon (grandfather of the historian),

* A. Lincoln, *Some Political and Social Ideas of English Dissent 1763–1800*, CUP, Cambridge, 1938, p. 272

when he engaged in pamphlet warfare with deists and latitudinarians (fence-sitting Anglicans who refused to commit to any definitive theological position), he settled at King's Cliffe near Stamford to live a life of simplicity and Christian community. With two female companions, Law gave himself to spiritual exercises and charity. They founded schools and almshouses and distributed alms to the poor.

Law took the mystic's path of confronting the world by withdrawal from it. That does not mean that he had nothing to say about the prevailing social mores. He was stringent in pointing out the indisciplined lives of those who espoused atheism or deism (for example, he wrote a pamphlet denouncing the loose morals associated with stage entertainments). His diagnosis of society's ills was as vigorous in words as Hogarth's was in paint. But he refused to enter into philosophical debate. 'Philosophical religion' he dismissed as a contradiction in terms. Christianity, he said, was about love or it was about nothing. For him religious belief and religious living were succinctly expressed in the words of the First Epistle of John: 'We love because God first loved us'.* He took up the challenge of rationalism in *The Case of Reason* in 1731. He demanded to know how, without revelation, we can say anything about God. The Creator must be infinitely beyond the reach of human reason. This is why God accommodates our limitations by showing himself in terms of what we can grasp naturally and primarily by coming among mankind in human form. The practical Christianity by which Law and his companions lived was a powerful bulwark for the theological and devotional principles he advocated in his writings.

The book of his that was destined to have the most powerful impact was *A Serious Call to a Devout and Holy Life* (1728). One of the first readers of this challenging treatise was the young Samuel

* 1 John 4:19

Johnson, lately arrived as a student in Oxford. He later told his biographer,

> I took up Law's *Serious Call to a Holy Life*, expecting to find it a dull book (as such books generally are) and perhaps to laugh at it. But I found Law quite an overmatch for me and this was the first occasion of my thinking in earnest of religion, after I became capable of rational enquiry.*

Law's rejection of reason as the means of laying hold of truth leads to the question, how can anyone gain access to the divine? His answer is 'conversion'. God has not hidden himself from man. On the contrary, revelation is constantly available through nature, Scripture and the inner light. It is man who has ceased to respond.

> As soon as the will of man turns to itself and would, as it were, have a sound of its own, it breaks off from the divine harmony and falls into the misery of its own discord.†

The only way to reverse this process is through death – death to the selfish half-life to which we cling – in order to be reborn. Then our spirit will point 'as constantly to God as the needle touched with the loadstone does to the north'.‡ All that man can do to effect this change, to experience God, is not to engage in philosophical speculation but, simply, to *desire* God – to desire him as a drowning man craves air. Law's advocacy of introverted, devotional discipline was one of the trumpet calls which roused many to a more active religious life. But there were others. In fact, during the decade 1733–1743, as a result of 'conviction preaching',

* J. Boswell, op. cit., p. 39

† William Law, *Collected Works*, 2013, VI, p. 27

‡ Ibid., VII, p. 139

there were dramatic outbreaks of religious revival in Wales, Scotland and Massachusetts, as well as in various parts of England.

In chronological terms the Great Awakening in America began first. The east-coast states had developed rapidly in the first century or so of their existence. In general terms they replicated the European societies from which they had 'escaped', mimicking the social structures, religious rivalries and cultural institutions of the Old World. They had their own centres of higher education, staffed by scholars who were on a par with and in connection with their European counterparts. One of the brighter students, who entered Yale in 1716, was Jonathan Edwards (1703–1758). He immersed himself in natural sciences and was particularly influenced by the works of Locke, whose *Essay Concerning Human Understanding* he valued more highly, as he said, than fistfuls of gold and silver.

The enthusiasm for observation and experimentation never left him and he ventured into print with various reflections on natural history. He found no conflict between the workings of nature and the Calvinist concept of God. Indeed, the more he studied the Creation, the more in awe he became of the Creator. For years he laboured intermittently at, but never completed, *A Rational Account of the Main Doctrine of the Christian Religion*. Following Locke and Newton he conceived of sensation as the root of all human ideas and God as the root of all sensation.

But he was always aware of the limitations of intellectual enquiry. As a committed Calvinist he was convinced that a sovereign God can only be known by those to whom he chooses to reveal himself. Edwards brought definition to the distinction between faith and reason, rationalism and revelation. It was an elucidation of what had often been urged before by Pascal, Calvin and Christian thinkers as far back as Augustine and St Paul. Seekers after truth, Edwards insisted, might formulate concepts of 'God' based on philosophical speculation and interpretations of the workings of the cosmos but religious conversion entailed a radical redirection of a person's thought processes. This could only be brought about by the work of the Holy Spirit, who originated 'a new inward perception or sensation of their minds'.* This was the evangelical response to the problem posed by Locke in *The Essay Concerning Human Understanding.*

It was during his student years that he wrestled with the 'What must I do . . .' question that had troubled countless others before him. Like many of them, he lived by an arduous regimen of devotional exercises and self-denial before making the breakthrough into 'assurance', the realisation that God *had* accepted him. He went on to become a lecturer at the university before seeking ordination to the Presbyterian ministry. In 1727, he became a pastor in the church in Northampton, Massachusetts. His preaching immediately proved effective. As he himself insisted, there was nothing new about the gospel of justification by faith that he presented, but he did so with such ardour that his congregation increased by three hundred in six months. The Awakening spread throughout much of New England, helped in no small measure by Edward's printed sermons. Many of the conversions were accompanied by scenes of hysteria. They were also attended by angry, even violent, opposition from Church leaders opposed to 'enthusiasm'. Both phenomena would repeat themselves in other areas where such revival occurred.

The eighteenth-century religious revival in Britain will always be associated with the names of the Wesley brothers and George Whitefield (1713–1770). John (1703–1791) and Charles (1707–1788) Wesley's ancestors were dissenters but their father, Samuel, had made his peace with the Establishment and accepted preferment as rector of Epworth in the rich Lincolnshire farmland.

* Cf. D.W. Bebbington, *Evangelicalism in Modern Britain, A History from the 1730s to the 1980s*, Routledge, Oxford, 1989, p. 48

For Samuel, social acceptance required involvement in the cultural and intellectual life of the region. This included membership of the Spalding Gentlemen's Society or, to give it its full title, The Society of Gentlemen for the Supporting of Mutual Benevolence and Their Improvement in the Liberal Sciences and in Polite Learning. There is more than a little significance in the foundation of this society in 1710. It was the first provincial debating club and part of the trend to introduce civilised discourse to the rural elite. The Spalding Society welcomed among its members, supporters and occasional speakers Sir Isaac Newton, Sir Hans Sloane (the president of the Royal Society whose collection later formed the nucleus of the British Museum), the Rev William Stukely (antiquarian and archaeological pioneer of Stonehenge and Avebury excavations), the poets Alexander Pope and John Gay and the antiquarian and leading engraver George Vertue. Samuel's children, thus, grew up in an atmosphere of earnest scientific and philosophical enquiry.

John and Charles were sent up to Oxford, set on the course to follow the family 'trade'. Both were diligent students as well as being disciplined in their devotional life. Their seriousness and disinclination to participate in the more boisterous activities of their fellows earned Charles and his small circle the derogatory name of 'Holy Club'. The roots of Methodism lay in the dedication of this group to an organised regimen of self-examination and service to others.

John shared the zeal of his brother but was the more intellectually gifted of the two. His prodigiously wide reading list embraced spiritual writers such as William Law and the leading philosophers of the age – Boyle, Locke, Newton and even Francis Bacon. John was essentially an activist, which was why he eventually turned his back on Law and the mystical tradition. He always maintained a lively interest in the workings of nature and the scholars who grappled with the understanding of it. Towards the end of his life he wrote,

> The immortal man to whose genius and indefatigable industry philosophy owed its greatest improvements and who carried the lamp of knowledge into paths of knowledge that had been unexplored before, was Sir Isaac Newton, whose name was revered and his genius admired, even by his warmest adversaries.*

Wesley was never an enemy of rational enquiry. He followed Locke in claiming experience as the basis of knowledge and, for him, experience came both from understanding of the natural world and putting to the test in his own life what he received from revelation. What he opposed was on the one hand the emotional excesses of 'enthusiasm' and on the other the unemotional aridity of rationalism that led to deism and atheism. His early years were spent in the quest for a faith that satisfied both head and heart, a faith that 'worked'. We might reasonably call him a 'spiritual empiricist'. His pilgrimage to this shrine led him along a winding path of intellectual enquiry and emotional upheaval.

Charles was also searching for enlightenment. In 1734 the brothers answered the call to go as missionaries to Georgia. Like their personal devotional rigours, this move was as much for their own spiritual comfort as for anyone else's. John candidly admitted as much: 'I hope to learn the true sense of the Gospel of Christ by preaching it to the heathen.' Whatever either of the brothers learned it was of no immediate comfort or positive effect. They were both back in England inside three years, leaving behind broken relationships and a great deal of ill feeling. However, on the outward journey they had spent time with a party of Moravians, and German Pietism provided the next contribution to their religious development. These people had something that the brothers lacked and knew they lacked. Soon after his return to England, John

* J. Wesley, *The Concise Ecclesiastical History from the Birth of Christ to the Beginning of the Present Century*, 1781, p. 332

wrote in his diary, 'I want that faith which none can have without knowing that he hath it.'* The story of how John attended a nonconformist meeting in London's Aldersgate Street and found his heart 'strangely warmed' (an experience shared by Charles a few days later) is well known. The brothers achieved that state of religious experience known as 'assurance', which incorporated but went deeper than intellectual conviction. John, with his typical thoroughness, set off almost immediately for Herrnhut, to learn more of the type of affective religion he had now embraced.

Meanwhile, the Wesleys' friend, George Whitefield, had begun his revivalist preaching career. This son of a Bristol wine merchant went to Oxford in 1732 and became one of the Holy Club 'methodists' in 1735. On leaving the university he was ordained and soon began to obtain a reputation as a preacher. In 1738, just as the Wesleys were returning, Whitefield set out for Georgia, where he took up a preaching and philanthropic ministry. He concluded that one of the most urgent social needs of the colony was an orphanage. Establishing, developing and funding this facility for the care and education of abandoned or fatherless children became one of the two aims to which he devoted the rest of his life. The other was preaching spiritual revival. He addressed crowds of people most often in the open air, on both sides of the Atlantic, dividing his time between campaigns in Britain and North America.

John Wesley initially found Whitefield's method of proclamation unseemly but he soon discovered that he, too, possessed his friend's gift of attracting and stirring crowds of thousands of people – and Benjamin Franklin once calculated that Whitefield could make himself heard and understood by an audience of thirty thousand. These two evangelists, each making horseback journeys of prodigious length and presenting their message to large crowds, were the most influential members of

* Cf. D. W. Bebbington, op. cit., p. 49]

the English-speaking world in the first half of the eighteenth century. Franklin also wrote of Whitefield that his influence wrought a remarkable transformation in the people of New England. 'From being thoughtless or indifferent about religion, it seem'd as if all the world was growing religious, so that one could not walk thro' the town in an evening without hearing psalms sung in different families.'*

Evangelical preacher George Whitefield and deist scientist and philosopher Benjamin Franklin were close friends for many years. One of Whitefield's initiatives was the establishment of a charity school for poor boys in Philadelphia. When funds for the project dried up it was Franklin who in 1749 persuaded the newly formed Academy of Philadelphia, a major torch-bearer of the Enlightenment in the New World, to buy the premises (comprising Whitefield's meeting house as well as the school) as its home. The deeds stipulated that on-site provision must always be made for the school. In 1753 a girls' school was also founded. Two years later, the College of Philadelphia came into existence on the same site. In the fullness of time these initiatives led to the foundation of the University of Pennsylvania, destined to be the training ground for many leaders of the United States of America. In Britain the emergence of Methodism was an equally significant agent of religious, social and (eventually) political change.

That coming together of religion, science and education in the service of a new nation might suggest a good place to end our discursive exploration of the numerous ways in which our ancestors of the sixteenth to eighteenth centuries tried to understand 'life, the universe and everything', and to make what seemed to them to be appropriate responses. However, it does leave aside one of the themes we have encountered. What about magic?

It was in 1736 that the British parliament passed an Act repealing all

* B. Franklin, *The Autobiography of Benjamin Franklin*, Houghton, Mifflin and Company, New York, 1888, p. 131

previous witchcraft statutes. It decreed that magic does not exist and provided that anyone claiming to wield occult powers would, in future, be liable to prosecution for fraud. Fifteen years later, a certain Thomas Colley was hanged in connection with a case involving witchcraft. But the condemned man was no witch – quite the reverse. His crime was stirring up hatred against two aged neighbours and inciting a mob who murdered the unfortunate couple. Clearly, statute law could define what was criminal but had no understanding of the origin of evil or how to eradicate it – subjects that continued to concern and enthral people.

In 1764 Horace Walpole published *The Castle of Otranto*, a book that inaugurated a new literary genre, the Gothic novel, whose mix of antiquity, romance, horror and occultism was but the latest manifestation of the ghost stories that had always fascinated people. Belief in the supernatural origin of evil is a very deep taproot in most cultures. John Wesley expressed the feelings of Christians of all stamps when he protested against rationalists who argued 'dogmatically against what not only the whole world, heathen and Christian, believed in past ages, but thousands, learned as well as unlearned firmly believe at this day'.* The culprits he had in mind were the fashionable materialists who rejected supernaturalism and made fun of religious revivalism. The most savage lampoon was a 1761 print by William Hogarth entitled *Credulity, Superstition and Fanaticism*. A complex satire showed Whitefield preaching to a congregation displaying various states of religious ecstasy and moral delinquency. Not content with 'exposing' Methodism, Hogarth also took some swipes at Catholicism. This was no rational and enlightened response to religious belief; it was the irrational and emotional reaction of materialism. The real debate between science and superstition continued – and continues. As author Roy Porter expounds:

* J. Wesley, *Journal*, ed. P. L. Parker, New York, 1906, III, p. 330

> What was happening was not the victory of light over darkness but a cultural revamping. Indeed, the paeans to progress once raised by philosophers and positivists, liberals and Whig historians, now seem highly problematic. With the benefit of cultural anthropology and the sociology of knowledge, today's historians, have abandoned the appealing but question-begging evolutionism that once celebrated the ascent of men from magic to science, from religion to reason; celebrating the triumph of truth has given way to analysing structures of belief.*

In truth, our story has no end, just as it had no beginning. If pursuit of *scientia* has led us anywhere it is to the point of knowing that we cannot know. The vastness of the universe and the complexities of our own humanity alike convince us that we are both doomed and privileged to go on exploring.

As we look forward in the hope of fresh understanding, so it is instructive also to look back to see how some of the great minds of the past approached the mysteries of being. That is all that has been attempted in the foregoing pages. To add an epilogue boldly labelled 'Conclusions' would be presumptuous in light of the vastness of our subject and the brilliant thinkers who have engaged in it. My task has simply been to introduce some of these intellectual giants and set them in their historical context, the better to appreciate them and their attempts to illumine the dark mystery of our existence.

And yet, I must come clean and confess another motive – not ulterior, since I more than hinted at it in the Introduction – I have tried to dispose of one superstition to which, whatever serious students now recognise, remains a common misconception. Not the belief in fairies

* R. Porter, 'Witchcraft and Magic in Enlightenment, Romantic and Liberal Thought' in M. Gijswijt-Hofstra, B. P. Levack and R. Porter, *Witchcraft and Magic in Europe*, A&C Black, London, 1999, V, pp. 193–4

or their modern counterparts, creatures from outer space. Not the stubborn religious instinct that preachers, prophets, poets, artists and musicians have always insisted on, which is that there is a purpose in our existence. But the belief in the all-sufficiency of the human intellect which, supposedly, solves more problems than it creates. That is a myth – and a dangerous myth. If ever we stop wondering about 'life, the universe and everything' and turn our thoughts inwards on ourselves, our spirits will implode; we shall become beings that exist merely because we exist, meaningless cyphers in a meaningless cosmos.

Alexander Pope wrote his most philosophical poem, the 'Essay on Man', in the years immediately prior to the revivals sparked by Whitefield and Wesley. The summation of his argument contains lines that may serve as a fitting coda to this study.

> Slave to no sect, who takes no private road,
> But looks through Nature, up to Nature's God,
> Pursues that Chain which links the immense design,
> Joins heaven and earth, and mortal and divine;
> Sees that no being any bliss can know,
> But touches some above and some below;
> Learns from this union of the rising whole,
> The first, last purpose of the human soul,
> And knows where Faith, Law, Morals all begin,
> All end in love of God and love of Man.

Bibliography

The literature on such a diverse subject is immense. This bibliography is simply a list of standard works by major thinkers and introductory essays that contain references for more detailed reading.

Allott, S., *Alcuin of York: His Life and Letters*, William Sessions Limited, York, 1974

Ankarloo, B. and Clark, C. (eds.), *History of Witchcraft and Magic in Europe*, vol. 4 & 5, University of Pennsylvania Press, Pennsylvania, 2002, 1999

Anscombe, E. & Geach, P. T. (eds.), *Descarte's Philosophical Writings*, Nelson's University Paperbacks, 1970

Aubrey, J., *Brief Lives*, A. Clark, ed. (1898), Penguin Books, London, 1972

Bacon, F., *Essays*, OUP, Oxford, 1921

Bacon, F., *The Advancement of Learning*, OUP, Oxford, 1974

Bainton, R. H., *Hunted Heretic – The Life and Death of Michael Servetus*, Beacon Press, Boston, 1960

Ball, P., *The Devil's Doctor, Paracelsus and the World of Renaissance Magic and Science*, Farrar, Straus and Giroux, New York, 2006

Barzun, J., *From Dawn to Decadence: 1500 to the Present* (1656), HarperCollins, London, 2000

Bebbington, D. W., *Evangelicalism in Modern Britain, A History from the 1730s to the 1980s*, Unwin Hyman, London, 1989

Bell, S. G., *The Great Fire of London in 1666*, Bodley Head, London, 1923

Berry, R. J. (ed.), *The Lion Handbook of Science and Christianity*, Lion Hudson Plc, Oxford, 2012

Boswell, J., *The Life of Samuel Johnson* (abridged), Dent, London, 1909

Bouwsma, W. J., *The Waning of the Renaissance, 1550–1640*, New Haven, Connecticut, 2002

Boyle, R., *The Works of Robert Boyle*, M. Hunter and E. Davis, eds., Pickering & Chatto, London, 1999–2000

Braggae, F., *A Full and Impartial Account of the Discovery of Witchcraft . . .* (1712), Gale Ecco, 2010

Brant, S., *The Ship of Fools*, trs. Alexànder Barclay, 1509

Brinkley, R.F., *Arthurian Legends in the Seventeenth Century*, Routledge, Oxford, 2016

Brooke, J. & Cantor, G., *Reconstructing Nature, The Engagement of Science and Religion*, Edinburgh, OUP, Oxford, 1998

Brown, C., *Christianity and Western Thought – A History of Philosophers, Ideas and Movements*, Vol. 1, Intervarsity Press, Illinois, 1990

Bunge, W. van (ed.), *The Early Enlightenment in the Dutch Republic, 1650–1750*, selected papers of a conference held at the Herzog August Bibliotik, Wolfenbüttel 22 –23 March 2001, 2003

Burr, G.L. (ed.), 'The Witch Persecutions', in *Translations and Reprints from the Original Sources of European History*, University of Pennsylvania Press, Philadelphia, 1898–1912

Calvin, J., *Avertissement Contre L'astrologie Qu'on Appelle Judiciaire*, Geneva, 1549

Caspar, M., *Kepler*, Dover Publications, New York, 1993

Chapman, A., *Physicians, Plagues and Progress – The History of Western Medicine from Antiquity to Antibiotics*, Lion Hudson, Oxford, 2016

Chapman, A., *Stargazers – Copernicus, Galileo, the Telescope and the Church*, Lion Books, Oxford, 2014

Chenu, M .D., *Toward Understanding Saint Thomas*, Regnery Publishing, Washingon, 1964

Cohn, N., 'The Non-existent Society of Witches', in M. Marwick (ed.), *Witchcraft and Sorcery*, Penguin, London, 1986

Cohn, N., *The Pursuit of the Millennium, Revolutionary Millenarians and Mystical Anarchists*, OUP, Oxford, 1957

Copernicus, N., *De Revolutionibus Orbium Coelestium*, trs. E. Roan, John Hopkins University Press, Baltimore, 1978

Cottingham, J., Stoothoff, R. & Murdoch, D. (eds.), *The Philosophical Writings of Descartes*, CUP, Cambridge, 1985

Cottreet, B., *Calvin – A Biography*, Westminster John Knox Press, Kentucky, 2007

Creham, F. J., 'The Bible in the Roman Catholic Church from Trent to the Present Day' in Greenslade, S .L. (ed.), *The Cambridge History of the*

Bible – The West from the Reformation to the Present Day, CUP, Cambridge, 1963

Dewhurst, K., *Dr Thomas Sydenham (1624–1689) His Life and Original Writings*, The Wellcome Historical Medical Library, London, 1966

Dickens, A. G., *Lollards and Protestants in the Diocese of York 1509–1558*, OUP, Oxford, 2012

Dunn, J., *Locke*, OUP, Oxford, 1984

Edmunds, D. & Eidinow, J., *Rousseau's Dog: Two Great Thinkers at War in the Age of Enlightenment*, Faber, London, 2006

Edwards, P. (ed.), *The Encyclopedia of Philosophy*, Macmillan, New York, 1972

Evans, R. J. W., *Rudolf II and His World – A Study of Intellectual History 1576–1612*, Clarendon Press, Oxford, 1973

Evelyn, J., *The Diary of John Evelyn*, ed. E. S. de Beer, OUP, Oxford, 1959

Foscarini, P.,'Epistle concerning the Pythagorian and Copernican Opinion of the Mobility of the Earth and Stability of the Sun . . . ' in *Mathematical Collections and Translations*, trs. T. Salusbury, 1661

Franklin, B., *The Autobiography of Benjamin Franklin*, Houghton, Mifflin and Company, New York, 1888

Fulbrook, M., *Piety and Politics, Religion and the Rise of Absolutism in England, Württemberg and Prussia*, CUP, Cambridge, 1983

Fuller, N., *Great Books of the Western World*, XXV, Encyclopaedia Britannica, Chicago, 1952

Galilei, G., *A Discourse Concerning Two New Sciences*, Dover Books, New York, 1954

Galilei, G., *Dialogue Concerning the Two Chief World Systems*, trs. S. Drake, University of California Press, Berkeley, 1953

Gawthrop, R., *Pietism and the Making of Eighteenth-Century Prussia*, CUP, Cambridge, 1993

Gijswijt-Hofstra, M., Levack, B. P. & Porter, R. *Witchcraft and Magic in Europe*, Vol. 5, The Athlone Press, London, 2001

Gurevich, A. J., *Categories of Medieval Culture*, Routledge, Oxford, 1985

Guthrie, D., *A History of Medicine*, Thomas Nelson, Nashville, 1945

Harrison, G. B., *A Jacobean Journal – Being a Record of Those Things Most Talked About During the Years 1603–1606*, Routledge, London, 1941

Harsnett, S., *A Declaration of Egregious Popish Impostures* (1603), EEBO Editions, 2010

Heywood, T., *The Wise Woman of Hogsdon* (1638), Scholar's Choice, New York, 2015

Hill, C., *Milton and the English Revolution*, Faber, London, 1977

Hill, C., *Some Intellectual Consequences of the English Revolution*, University of Wisconsin Press, Wisconsin, 1980

Hobbes, T., *Works*, Clarendon Press, Oxford, 1984

Huizinga, J., *The Waning of the Middle Ages – Life, Thought and Art in France and the Netherlands in the Fourteenth and Fifteenth Centuries*, Edward Arnold, London, 1967

Hume, D., *Enquiries Concerning Human Understanding and Concerning the Principles of Morals*, ed. L. A. Selby-Bigge, OUP, Oxford, 1975

Hume, D., *The Natural History of Religion*, ed. H. E. Root, A&C Black, London, 1956

Hyde, E., *A Brief View and Survey of the Dangerous and Pernicious Errors to Church and State in Mr Hobbes's Book entitled Leviathan*, 1673

Jaki, S. L., *The Ash Wednesday Supper by Giordano Bruno*, University of Chicago Press, Chicago, 1969

James IV of Scotland, *Daemonologie, In forme of a dialogue*, 1603

Janowski, Z., *Cartesian Theodicy: Descartes' Quest for Certitude*, Springer Publishing, New York, 2000

Jensen, H. J. (ed.), *The Sensational Restoration*, Indiana University Press, Indiana, 1996

Jones, W. T., *Masters of Political Thought: Machiavelli to Bentham*, Vol. 2, George G. Harrup, London, 1960

á Kempis, T., *The Imitation of Christ*, trs. W. Benham, [n.pub], 2016

Kuhn, T., *The Copernican Revolution – Planetary Astronomy and the Development of Western Thought*, Harvard University Press, Harvard, 1966

Lambert, M., *Medieval Heresy*, Wiley-Blackwell, Oxford, 1992

Latham, R. G. (ed.), *Epistolae responsoriae*, 1848

Latini, B., *Livre de Tresor*, Gondoliere, Venice, 1839

Law, W., *Collected Works*, 2013, Kindle edition

Leibniz, *Reflections on the Common Concept of Justice*, c. 1702

Lilly, W. & Ashmole, E., *William Lilly's History of His Life and Times from the Year 1602 to 1681*, Harvard University Press, Massachusetts, 1822

Lincoln, A., *Some Political and Social Ideas of English Dissent 1763–1800*, CUP, Cambridge, 1938

Locke, J., *An Essay Concerning Human Understanding*, ed. P. H. Nidditch, Clarendon Press, Oxford, 1998

Locke, J., *Works*, OUP, Oxford, 1975

Luther, M., *The Bondage of the Will*, trs. Packer, J., Johnston, O., James Clarke, Cambridge, 1957

Luther's Works, eds. H. T. Lehmann, & J. Pelikan, Concordica Publishing House, St Louis, Missouri, 1955

MacCulloch, D., *Reformation – Europe's House Divided 1490–1700*, Penguin, London, 2003

Macfarlane, A., *Witchcraft in Tudor and Stuart England*, Routledge, London, 1970

Markus, R., *The End of Ancient Christianity*, CUP, Cambridge, 1990

Martinich, A. P., *Thomas Hobbes*, CUP, Cambridge, 1997

Marwick, M., *Witchcraft and Sorcery: Selected Readings*, Penguin, London, 2010

Mebane, J. S., *Renaissance Magic and the Return of the Golden Age*, University of Nebraska Press, Lincoln, 1989

Milano, P. (ed.), *The Portable Dante*, Penguin, London, 1977

Mossner, E. C., *The Life of David Hume*, OUP Oxford, 1980

Nauert, C .G., *Agrippa and the Crisis of Renaissance Thought*, University of Illinois Press, Illinois, 1966

Nauert, C .G., *Humanism and the Culture of Renaissance Europe*, CUP, Cambridge, 2006

Oldridge, D., *The Devil in Early Modern England*, Sutton Publishing, Stroud, 2000

Pachter, H. M., *Paracelsus – Magic into Science*, Henry Schuman, New York, 1951

Paget, S., *Ambroise Paré and his Times*, G. P. Putnam, New York, 1897

Palmer, R. R., *The Age of Democratic Revolution: The Challenge*, Princeton University Press, New Jersey, 1959

Pascal, B., *Pensées*, trs. A. J. Krailsheimer, Penguin, London, 1966

Pascal, B., *Thoughts, Letters, Minor Works*, Harvard Univeristy Press, Massachusetts, 1910

Pollard, A. W., *Records of the English Bible*, OUP, Oxford, 1911

Popham, A .E., *The Drawings of Leonardo da Vinci*, The Reprint Society, London, 1952

Price, J., *Danmonii Orientales Illustres Or The Worthies of Devon*, Rees and Curtis, London, 1810

Reay, B. (ed.), *Popular Culture in Seventeenth-Century England*, Longman, London, 1988

Redwood, J., *European Science in the Seventeenth Century*, David & Charles, Exeter, 1997

Root, F. K. (ed.), *Lord Chesterfielde's Letters to His Son and Others*, Dutton, New York City, 1957

Rousseau, J. J., *Emile*, trs. B. Foxley, Dent, London, 1921

Ruickbie, L., *Faustus – The Life and Times of a Renaissance Magician*, The History Press, Stroud, 2009

Rupp, G., *The Righteousness of God – Luther Studies*, Hodder & Stoughton, London, 1953

Ryan, W. G. (trs.), *The Golden Legend: Readings on the Saints*, Princeton University Press, New Jersey, 1993

Smith, P., *Erasmus – A Study of his Life, Ideals and Place in History*, Frederick Ungar Publishing Co., New York, 1962

Spedding, J., *The Life and the Letters of Francis Bacon*, Longman, Green, Longman, and Roberts, London, 1861

Spellman, W. M., *John Locke*, Macmillan, New York, 1997

Stark, R., *The Victory of Reason, How Christianity Led to Freedom, Capitalism and Western Success*, Random House, New York, 2005

Summers, M. (trs.), *Malleus Maleficarum*, Pushkin Press, London, 1951

Thomas, K., *Religion and the Decline of Magic: Studies in Popular Belief in Sixteenth- and Seventeenth-Century England*, OUP, Oxford, 1971

Thoren, V .E., *The Lord of Uraniborg – A Biography of Tycho Brahe*, CUP, Cambridge, 1990

Thornton, T., *Prophecy, Politics and the People in Early Modern England*, Boydell Press, Woodbridge, 2006

Vesalius, A., '*De Fabrica Corporis Humani*' (1543), Preface in *Proceedings of the Royal Society of Medicine*, Longman, London, July 1932

Wagner, R. & Briggs, A., *The Penultimate Curiosity – How Science Swims in the Slipstream of Ultimate Questions*, OUP, Oxford, 2016

Weber, E., *Apocalypses, Prophecies, Cults and Millennial Beliefs Through the Ages*, Random House, Toronto, 1999

Webster, C., *From Paracelsus to Newton, Magic and the Making of Modern Science*, CUP, Cambridge, 1982

Wedberg, A., *A History of Philosophy*, Vol. 3, Clarendon Press, Oxford, 1984

Wedgwood, C. V., *The Thirty Years' War*, Jonathan Cape, London, 1947

Weinberg, S., *To Explain the World – The Discovery of Modern Science*, Allen Lane, London, 2015

Wesley, J., *The Concise Ecclesiastical History from the Birth of Christ to the Beginning of the Present Century*, 1781

Wesley, J. *Journal*, ed. P. L. Parker, New York, 1906

West, G., *Observations on the History and Evidences of the Resurrection of Jesus Christ*, fourth ed., (1749), Gale Ecco, Michigan, 2010

Westfall, R. S., *Science and Religion in Seventeenth-Century England*, Yale University Press, New Haven, 1958

Wojcik, J. W., *Robert Boyle and the Limits of Reason*, CUP, Cambridge, 1997

Yates, F., *Giordano Bruno and the Hermetic Tradition*, University of Chicago Press, Chicago, 1964

Index